Mathematical predictive model

The Equations that replaces God: How the Cosmos works…

Timoteo Briet Blanes

Twitter: @TimoteoBriet
Mail: racecarsengineering@gmail.com
LinkedIn: www.linkedin.com/in/timoteobriet

➔ *Preface*

In the life of any person, you can do 2 things:

- Study or Believe

I have chosen the first option.

If you "believe" in Astrology, or that the Earth is Flat, or that Man has not gone to the Moon, or you are an anti-vaccine, or simply don't want to learn and advance, ***DON'T READ THIS BOOK, AS IT WILL MAKE YOU DOUBT YOUR "BELIEFS" AND YOU ARE IN DANGER OF KNOWING HOW THE "REAL" WORLD WORKS.***

If you want to remain a complete Ignoramus and live by your Beliefs, don't read this book. It is very important.

This Book-Theory, are various Articles explaining the full Theory that I want to show: a new vision of the Cosmos and its dynamics in all senses.

Approximately weekly, I will have new versions of these Articles, which I will post on Linkedin and Twitter mainly. I will be adding, modifying and improving each Article, completing and clarifying the concepts and examples.

➜ *Abstract*

Human beings have always felt the need to know and predict events future, using Science. Many natural phenomena are explained by numerical models; but there are as many numerical models as there are phenomena; this is problem, a big problem; ideally, many events should be explained using for this, the least amount of mathematical models.

Navier Stokes equations have been used for many years to simulate fluid dynamics. There are many particular cases in which these equations describe the dynamics of various phenomena as different as economics and meteorology. Most attempts to use these equations in a variety of fields, lie in properly defining the variables involved, giving them a physical explanation. This is an exciting challenge, of course and that is the main goal for this article: applying these equations, for explaining the Cosmos dynamic.

If it looks at many natural events, it will see that they evolve as a fluid: flocks of birds, vehicular or pedestrian traffic, are typical cases of analysis, but we can also observe this dynamic in events such as the stock market, the economy or even human relations.

Writers use language to produce tears, fear, longing and sensations in general; but even there are no words to express certain things; mathematicians use another language to explain phenomena and try to create new symbols to explain the most complicated phenomena.

If you learn Italian or Russian, you can talk to an Italian or a Russian; but if you learn Mathematics, you can talk to anyone, to Nature or to the Cosmos; it is a Universal language; it is even the only language in which anyone can invent words, structures, theories, etc. I am still learning words in Mathematics that allow me to understand the Cosmos.

When trying to explain any phenomenon, a "normal" person will describe the phenomenon in words, explaining the reasons for its existence, its evolution and perhaps even the principles on which it is based; a mathematician will explain the same phenomenon in a different language, using mathematical expressions. I, as a Mathematician, would like to express with my "language", envy, love, grief, loneliness or longing; I would like to know it in a short time....

However, we mathematicians also cry when we know how to explain a phenomenon....

All Scientists, for many years, have wanted to develop a unitary and general theory that explains all the phenomena of the Universe, from the largest to the smallest phenomena in the Universe. This work is a further contribution to that desire of science.

Most mathematical models offer unique solutions; what is intended in this work is, like the Schrodinger's equation to show solutions with a probability scale; that is to say: a particle will tend in a direction, with a given probability.

ARTICLE 1

The Mathematical Universe-Particles

→ *Introduction*

Cosmos (Carl Sagan): "Cosmos is all in the past, present and future."

All the matter in the Cosmos seems to behave in the same way: its dynamics is based with the same or similar laws, whatever the scale (not quantum scale....).

If it want to explain how the cosmos works, it is necessary:

- To detect patterns in events.
- To detect patterns between events.

And what is more important and transcendental:

- To detect patterns between numeric models that describes the events.

It will find similarities between phenomenon's and numeric models. This is one of the main goal.

We mathematicians speak a language. In language made up of words, groups of words, sentences, paragraphs and even feelings.

Mathematics can also be applied to solve physical problems, simulating "analogies"; for example, the case of traffic jams in cars, using the analogy of a water hammer in pipes: a multi-lane road can be installed to mitigate a traffic jam, in the same way that a water tank is installed, to reduce or eliminate water hammer (water tank and multiple roads, simulating a "expansion").

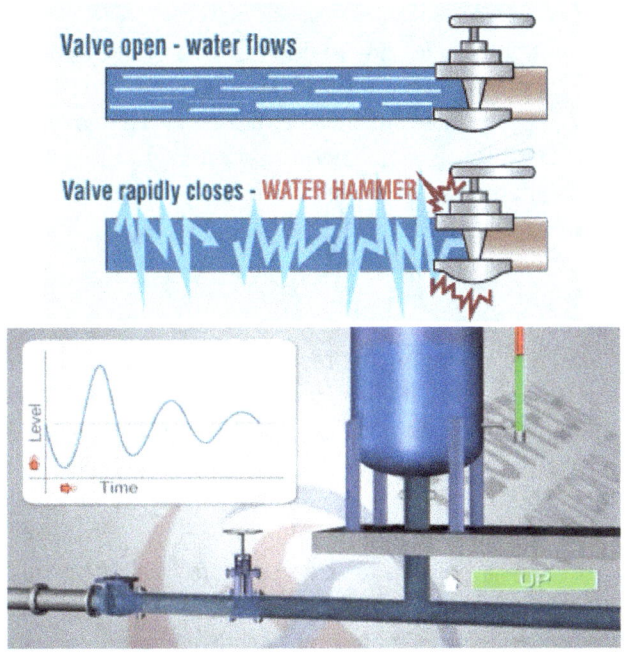

In a particle group and its movement, there is a force that push every particle; this force is the result of several forces (friction, magnetic, pressure, Coriolis, etc), which, acts on the particle. It is incredible, but this occurs in each of the particles of a group, and each of them, completely unaware of the forces and directions of other particles. This mutual ignorance perhaps is the main reason in the generation of precious figures or geometry and global structures.

The main goal for any mathematician is create numeric models about nature phenomenon. For that, is necessary discovery (or create artificially) patterns, and if it is possible lineally, but that, is not easy, and normally not real. As a writer, a mathematician thinks with languages and as all language have their rules, their pretty rules.... It is very nice is front a white paper and write ideas and translating dreams....

The human being expresses himself with numerous languages in the many places of the world where Mathematics is another language, with its rules. Any language serves to express feelings and facts; if the Mathematical language is capable of having words that express many feelings under many hypotheses or contexts, it will be able to express more faithfully or really, a natural fact. If Mathematics is not capable of expressing a phenomenon, it simply means that language must be advanced and improved until it is capable of modeling it.

A Mathematician therefore explains and describes "Reality" with a characteristic language. There are phenomena in Nature that are not yet explainable or easily explained. This is because there are not yet any mathematical words or sentences for it. The work of a good mathematician is double:

- To know how to use all the mathematical language that already exists.
- To create new words to explain phenomena.

For example, the continuity equation or divergence equal to zero, means that the density does not vary.

We, as human beings, have had the feeling on many occasions, of wanting to express certain feelings, ideas or dreams, but we cannot: there are no words that can express them.

We use then, unions of already existing words that normally express badly what we want to express. The fact of call one event as unpredictable is to assume ignorance. The goal so, is know the evolution in any coordinate of any object or event, from similarities. In the nature, there are a lot of think very weird, about patterns and data series:

- Benford law, applied for example, in distances of galaxies from earth ([31] Timoteo Briet Blanes): (brown=Benford law, yellow=data 4.000 galaxies data founds from Internet):

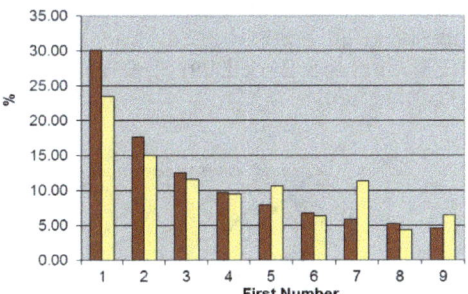

There are many fields where this Benford law applies, such as fraud detection (invoices, etc...). But also in other fields such as brightness in objects captured by the Fermi telescope, or other cases:

Natural law?

Under Benford's law, the first digits of the numbers in a set have a non-random distribution. Such sets are being discovered throughout nature

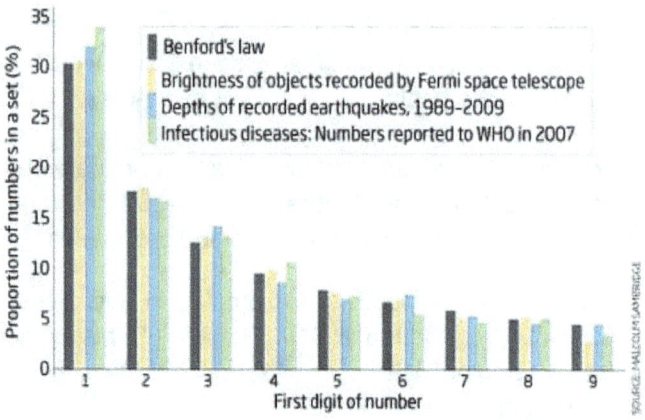

- Generation on Lissajous curves from different cases: for example in lift against position in a vibrating wing, etc… ([27] Timoteo Briet Blanes)→ 2 Articles about in the end of this work:

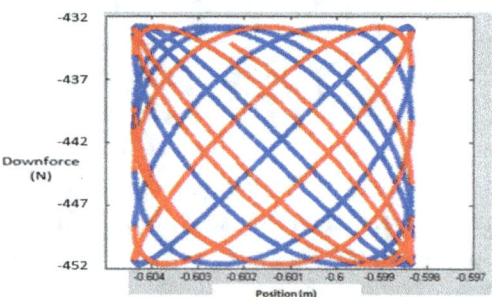

There are lot phenomenon's in Nature, with the "same" phase space as a Lissajous curves.

- Even, Vortex Street is sometimes formed when the wind from a star flows past a neutron star companion…. And also, Vortex Karman street, in different scales as a turbulences in cylinders, atmospherics events, tail in striped galaxies, etc….

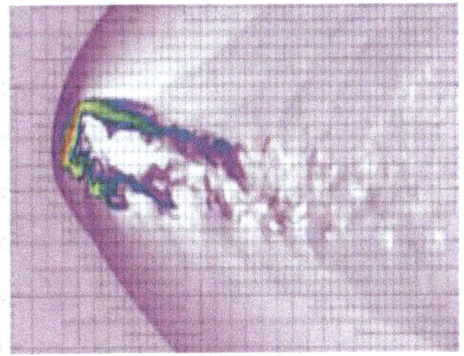

- It´s possible to see the vortices-disturbances created from the Kelvin-Helmholtz effect, also in different scales: clouds, Orion Nebula:

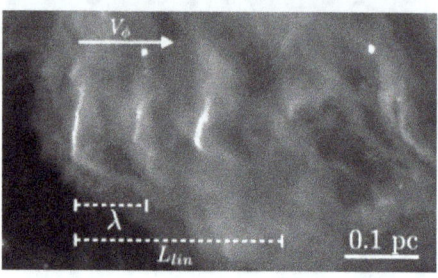

Many times, it is not easy to observe a well-defined or clear pattern. It´s therefore necessary to analyze data or geometries, perhaps indirect to the phenomenon, which may give rise to a possible pattern.

For example, in a Meteorites rain (Quadrantides – temporal data from J. M. Trigo - January 1992), it´s possible detect and analyze one geometry multifractal (may be because there is a random variable....); it's possible to create a graphic in 2D, with "d_i" the detection instant of meteorite "i":

➔ $d_i - d_{i-1}$ against $d_{i-1} - d_{i-2}$

([9] Timoteo Briet Blanes).

And also in structure of our galaxy: ([16] Vicent Martínez García).

And more: it's possible to see some phenomenon or properties as a fluid, in objects or particles dynamic.

For example, Bernoulli principle in accumulation-aggregation or exit of people from sport stadium, also sheep out of a stable, even it is possible apply fluids theories in vehicles traffic in cities, etc....

About the phenomenon prediction, if there are few laws which define him, it will be more complicated to know the evolution (chaos essence).

From all that is necessary to ask us, if there some think common for all these cases, some law able to means these examples.

That is the main goal for me: know how the nature think and decides, and create language or concept mathematics, pretty and simple, in order to explain any event, as a fluid or as a particle.

To know the evolution of any event means the introduction may be of a probability of to be or not to be. That is very important.

Can you fight the flow of a brave river trying to reach the edge? Surely it will be useless, but you can try....

Every person has his own will and is able to choose his destiny or displacement as a decision or choice, but the group dilutes that will; It might even alter your environment, but only the environment ....

A person solitary, is unlikely to originate or produce a "different" evolution of the whole; but it will be able to do so, only in the case of being able to generate a great impact that affects many people: the union, it makes force. When one speaks of "power," power is the ability to influence large numbers of people. The birds, don't know what is the geometry of a flock, but hi flights and moves....

Who, when a very dear one has died, has not thought that the world is going to stop, that the sun will not come out any more, or that everything will change, or that he will telephone on your birthday to congratulate you. Really the sun does not come out the same way and with the same beauty, but the world follows, and despite what happened, everything remains the same.... and never phone.... I need understand the Carl Sagan cosmos, but I and my actions, are very and quite insignificants....

There is a special relation between sloth or minimum energy principle and fluid dynamics: If I must to go from here to there, yes; I will go: but, with the minimum energy,,,,, If it ask question about universe, it would be able to understand it....

Some variables found in the Navier Stokes Equations have been adapted to analyze problems of accumulation of people in premises, as in ([8] Kazunori Shinohara and Serban Georgescu) without offering a clear and well-defined protocol for analyzing any problems related to pedestrians; in fact, in other articles on pedestrians, a protocol is drawn up but adapted to each problem, without defining a generic one. Furthermore, there is no discussion of the possibility of working with these equations, in "n" dimensions greater than 3.

In ([3] Jakimowicza and J. Juzwiszynb), the possible spirals or vortices that would be formed in problems are commented economic, if studied in 3 dimensions; but they do not offer either a clear and well-defined, for all kinds of economic problems. Same as above, nor is there any discussion of the possibility of working with such equations, in "n" dimensions greater than 3.

In short, all the articles and research I have been able to obtain and analyze have the same problems or process:

- They offer a numerical model and/or protocol of action, adapted to each problem.
- They do not work with Navier Stokes Equations in dimensions greater than 3.

What is attempted in this Article is to solve both problems.

First, the variables used in the numerical models are defined; then the methods by which matter accumulates are studied, to then describe numerical models for the dynamics of a group of particles; finally, the Navier Stokes Equations are analyzed, proposing an application protocol to describe and predict events, in general.

We are wrong describing the entire universe? For example, in Schrodinger or Navier Stokes equations, it´s possible to find derivatives order 2 as max; but, the nature work as that? Is missing derivatives of 5 order for example? Complicate to know that. In the flock birds of fluid case, every particle know the movement of the particle around her; if there are dependence of 5 order of derivatives, that would mean that the dependence is bigger (more fare away, not only around)....

→ *Definitions*

- *Event*

It cans considerer an Event, as any concept, dependent of time: Event, Phenomenon, Particle, Success, is the same concept; as Carl Sagan understood it: "all that has been, is and will be". A fluid for example, is a group of particles.

There are 2 types of events:

- Continuous events: those events that at any moment have a value. For example, the price of oil; in this case, a price increase is established, so that as long as the price does not vary more than this increase, the unit of discretion does not vary either.

- Occasional events or discrete: those are not continuous. For example bomb explosion, volcano eruption, etc.

- ***Dependence, dimension and events representation***

A coin is thrown: what is the probability that it comes out face? The answer seems pretty obvious. But, and if it is known that previously the same coin has been launched 50 times and has always has face? The answer is no longer so simple, besides that there are some explanations mathematically (Markov chain, etc....). Also analyze Bayes, Pascal and Anchenwall. Does it therefore influence what is known a priori of an event in order to predict it? Does knowledge influence? Yes that influences, indeed: if you ask us if it's going to rain an hour, just look at the sky and know if there are many clouds....

If an experiment is measured, it can affect the development of that experiment; let us suppose the following situation: we roll a die; what's the probability of it being a "6"? And if we know that before, it has been thrown

50 times and a "6" has always come out? We will answer the same thing? The previous knowledge, influences this test?

Given a dice, there is the same probability that any number from 1 to 6?

No.... ???? Each number is defined on each face, with hemispherical holes or part of hemispheres. These holes remove material from the dice, so that the face of number 6, for example, weighs less than the face of number 1 or any other...

It would be the same with the lottery balls, since each number is marked with paint, which, makes the more paint, the more aerodynamic drag the ball? Or marked with holes: more or less weight in every face. I'm very sure of that... I would like to test in CFD that...

It can see some about the probability "rare": in ([40] Joseph B. Keller) and [41] J. Strzalko, J. Grabski, A. Stefanski, P. Perlikowski and T. Kapitaniak): in the toss of a coin, without dropping it on the ground, the probability that the top of the coin when tossed is greater than the bottom; a probability of 0.52. In fact, the randomness, assuming 0.5, is derived from the bounces, if any, on the ground.

The glass is a material called "Amorphous":

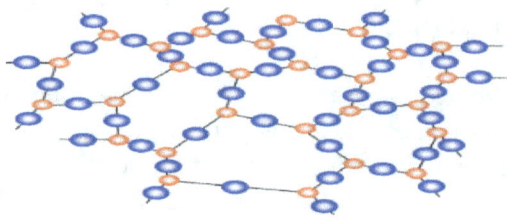

This material has the property of not transmitting normally, a vibration. In fact, is possible think about the glass, as a material with a viscosity very high....

Be 2 events; it is assumed that one of them varies and it is observed that the other event also varies or responds to the variation of the first. Are both events therefore dependents? One could say yes, as long as these mutual variations are known over a suitably long time, since, perhaps, the second event varies "coincidentally"....

A group of events can be represented by their relations between them, in the following ways:

- Springs, dampers, shock absorbers, fix bars or nothing, even combinations between.

- Fixed bar (positive or negative): one event moves in the same proportion as another to the same direction.
- Spring: it is defined analogously to the bar, but with a force of repulsion or attraction, as a spring.
- Damper: it is a displacement damper, applicable to bars and springs. Is a try to enter the variable "time" and velocity.
- Inerter. Is a try also, to enter the variable "time" and acceleration.

- Nothing: if are not dependents.
- Combinations between; for example: spring-damper, damper-inerter, etc.…
- Etc.…

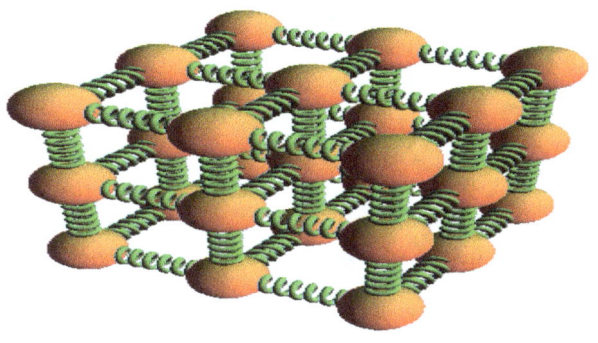

It is possible to apply "mass" (size) to the event, as "importance" or "transcendence" or "weight".

The options, therefore, of connections between events, are endless. All these relationships can work under linear and non-linear functions.

- An event is represented according to different "Coordinates", which are the variables on which the Event depends. The "Dimension" of the event is defined as the number of variables on which it is possible to represent it.

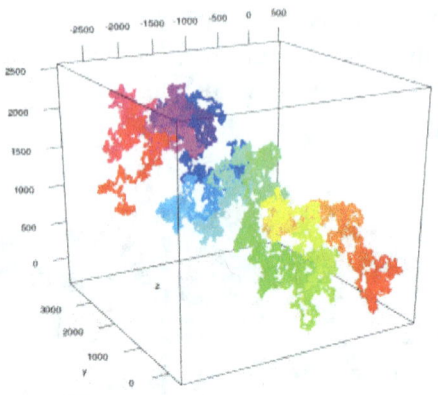

In the face of the evolution of an economic crisis, it always   asks   us: "until when?".

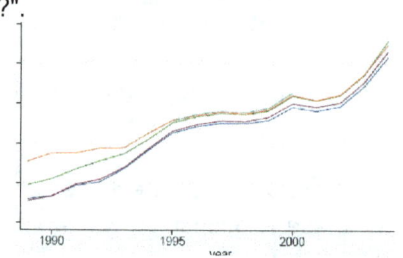

It do not know at all, when it will stop downloading, or when it will stop uploading in your case; but one thing is clear:

at some point it will stop going down.

There is nothing that goes up or down forever; like a diver, no matter how deep the waters you dives, "always" there will be a time when you touch the bottom or reach your maximum depth.

To say that the economy rises and falls alternately, like a saw tooth, is to admit our ignorance of how it evolves; besides, if he did not do it, it would be absolutely incredible to go up or down constantly ... Sure it would be surprised.

And another question:

Is there any merit in "leaving" that some stones, thrown into the sea, reach the bottom, is there merit in saying   that   they   will   reach   the   bottom?

15

Imagine a pool like an ocean; if we open the drain, sooner or later, it will empty...

The question always arises: "What to do?".

All governments "try" to mitigate the effects of the crisis, "doing things" under the options and criteria, more or less successful, that mark or govern their ideologies.

But also, it can all verify that these actions either have no appreciable effect, or are slightly appreciable in the very long term. If indeed it can see some effect, it is simply because the previous diver was already close to the bottom....

The world economy or global dynamics is the one that always prevails; it's like wanting to empty the sea, from glass to glass.

It´s true that before a small action, as it is to cover the drain of the ocean, we make it never empty; but we will know that it is not going to be emptied, in a very long time.

A bird knows only the movements of those around it; it is completely unaware of the rest; but the whole group moves forward and creates beautiful figures; if one bird alone tries to vary the movement of the whole flock, it will be extremely difficult; this will be achieved by the alteration of many birds.

It´s more: there are actions that do not affect "absolutely" in anything; therefore, it has 3 possibilities:

1. Do something and see its possible consequences in many and many years.
2. Do something that does not affect anything (and people see that something is done).
3. Let the global dynamics prevail and flow...

What is the best choice? The 3; At least, let's dedicate ourselves to enjoyment and that other rights are not affected. Sup 3 events (A, B and C); "A" fixed; then if "C" moves, "B" will move; but the greater "b" and the smaller "c", keeping "a" constant, the displacement of "B" will be less.

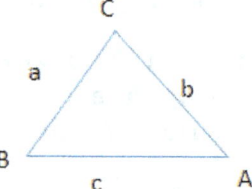

It is an example to observe that although we have 3 dependent events, certain displacements of one of them, may have very little importance on the others.

Any government that takes credit for taking a country out of a crisis lies: it simply has been lucky to be at the right time.

It can define "being alive" to that substance that is able to have notion or consciousness of the passage of time.

It is possible to perceive time in a different way; in fact, when it is sleeping or when it is older, it does so.

It´s time the necessary variable for there to be a dynamic? If everything were causal, the existence of time would not be necessary, since "everything" would already be defined and marked until eternity. It is also true that, as we have already seen, in the dynamics of a set of phenomena, only one of them lacks the power to modify fully; it is the randomness that marks this effect or influence.

Randomness is necessary in the universe, for whatever reason, but it is necessary.

In fact, let's think of 2 different phenomena (water flow and galaxy formation): time scales and time are different; the time is other coordinate so.

It is as if the dynamics of the universe invite us or force us to standardize time and its scale, in order to be able to compare.

It defines dimension of event, as a coordinates number of event; that is: every coordinate is a factor which the event depend (in our analysis or case). From these coordinates, it´s possible so, representing the event. That is, more or less, the Phase space.

ARTICLE 2

Definitions: Velocity, Density, Pressure, Compressibility, Temperature, Instability, Viscosity, Similarity Number, Tension and Expansion

- *Velocity "V"*

Velocity of an event "E" in a coordinate or direction "d" is defined as the number of events with respect to "d" ; that is:

$$\vec{V} = \frac{\partial E}{\partial \vec{d}}$$

The "d", can be the Time; then, the velocity have the dimensions "traditional" or "normal", that we know.

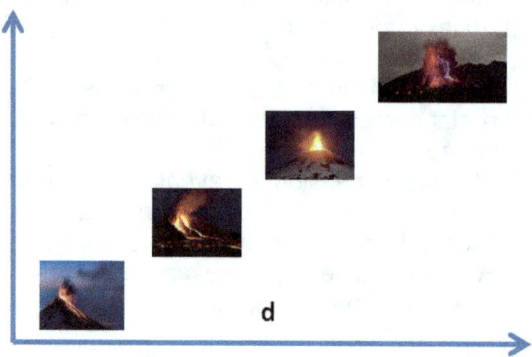

-   ***Density (ρ)***

In general, Density "ρ" in a direction "d", is defined as the quotient between the number of particles "NE" in a direction "d" enclosed in a ball of determined radius "R" and center of particle, and the volume of the ball. This definition is extended to "n" dimensions, defining the volume of a ball of "n" dimensions as:

$$\rho_{\vec{d}} = \vec{NE}_{d} / \frac{\pi^{n/2} R^{n}}{\Gamma(n/2+1)}$$

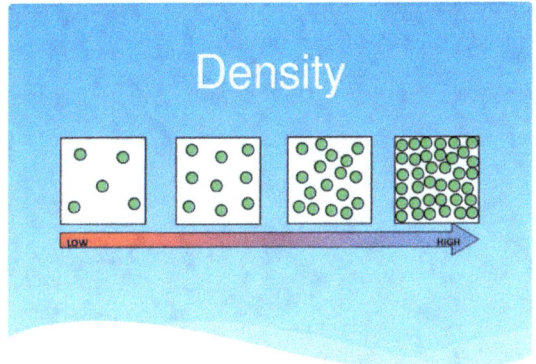

"z" is an integer and "Γ" being the Gamma function:

$$\Gamma(z) = \int_{0}^{\infty} t^{z-1} e^{-t} dt$$

This volume is greater and lesser against dimension; there is a max as a dimension (radio equal 1):

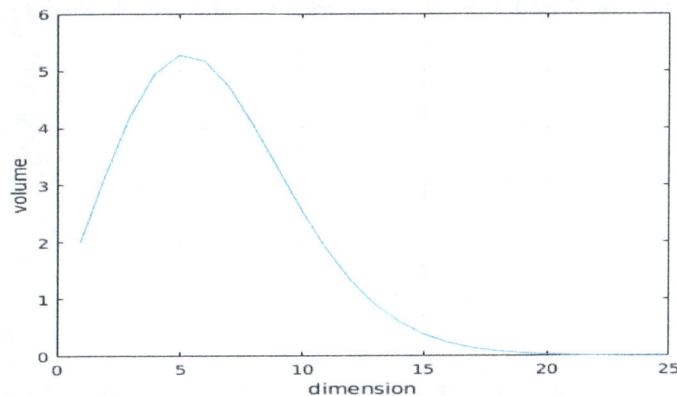

The time will be the most used dimension, but density can be calculated in any coordinate or direction.

If the density is a number against "Time", may be called this density, as a Frequency.

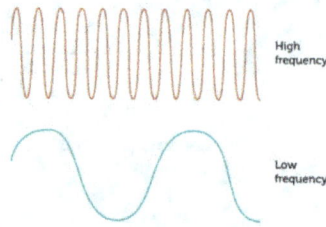

There are other types of density in space, which also influence the dynamics of particles; these densities can be grouped into the so-called Energy Density:

- Electro-Magnetic Density.
- Gravitational Density.
- Others, may be….

- **Pressure (P)**

First, it can think about pressure "P" in a direction "d" as a definition in Kinetic theory of gasses (proportional to "m" total mass particles, "1/Vol" volume in any dimension ("Vol"), "u" average velocity of particles in direction "d", impulse (m*u) and "N" number particles in a direction "d"); "P" is also a vector, because the velocity and density, also they are:

$$\vec{P} \propto \frac{m\,\vec{N}\,\vec{u}^2}{Vol} = \vec{\rho} * \vec{u}^2 * \vec{N} \propto \vec{\rho}\,\vec{u}^2$$

This concept is very important in galaxies formation and evolution or in general in cosmology. In this case, "P" is called "Ram Pressure", and very similarly, to

Einstein equation (simplified) E=mc² (m←→ρ) ("c" speed of light, "m" mass, "E" energy):

$$E = m\,c^2 = \frac{m}{Vol}\,c^2\,u \propto \rho * Vol$$

$$E \propto P * Vol$$

Let "Pe" in motion quantity ("v" velocity): Pe=mv; sup. 1 dimension "x"; "m" constant but "v" depends in every point; so:

$$\int \vec{Pe}\,dx \cong \frac{1}{2}m\,\vec{v}^2 = \vec{E}$$

Pressure is a measure of energy, as is the Quantity of Motion (=mv); in this case, an attempt is made to "normalize" the measure of mass, transforming it into mass per unit volume: density.

$$\vec{P} = \frac{\vec{Pe}}{m}\,\rho\vec{V} = \vec{Pe}\frac{\vec{V}}{vol}$$

- **Compressibility (Z,β β)**

It can have a fluid with compressibility but is necessary to know the velocity for this compression or expansion. This value is the divergence of velocity; that is: the variation of volume, and may be positive or negative. It suppose that events group, may be different pressure against the time or other variable. That is: 2 events in a fluid (as a particles set) non incompressible:

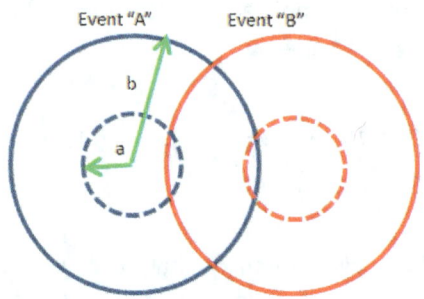

The 2 events "A" and "B", cannot be less than a distance "a" or more than a distance "b". In case of being more than "b", the events can be considered independent, in the first phase. These distances "a" and "b" can are different depending on temperature, pressure and density, for the same fluid-group of events (depends of a fluid type). The speed of compression and dilatation ("a" and "b", is function of spring-damper system, or other combination between forces, velocity, acceleration, etc...). The density of a fluid formed by particles depends directly on the compressibility and vice versa; compressibility is defined as the force applied to 2 particles to bring them closer together. Be a closed box full of billiard balls; if it tries to move the balls, it will be absolutely impossible.

But if there is some kind of compressibility, the balls will tend to move and pass one another.... (Tennis ball, for example).

From this reasoning, any people can perfectly understand Pascal's principle, or the transfer of forces between communicating vessels.

Let us suppose circular particles (2 dimensions); in an enclosure with such particles, there will be a certain pressure and a certain density; there could be another enclosure with other particles of a larger radius, with the same density, but the pressure will also be higher: the pressure depends on the size of the particles. If a particle "A" at the top of the cylinder is pushed, this force is distributed to all the particles around it, reaching the top of the other cylinder:

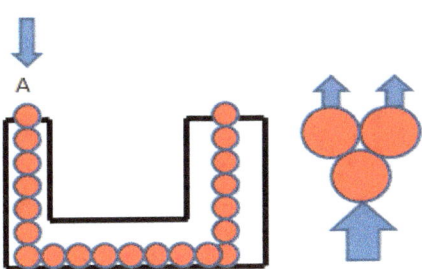

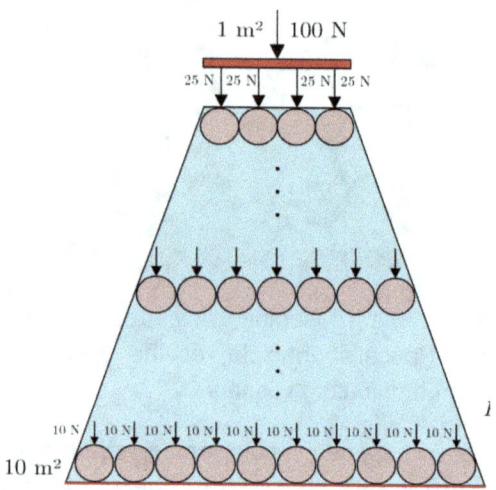

The compressibility can be defined also by:

- By "Z": "Vol$_m$ "is a molar volume, "R" is a fluid constant "T" is the temperature:

$$Z = \frac{PVol_m}{RT}$$

"R" depend of "a" and "b" (size movement particle).

- By "1/β" ("Vol" is a volume):

$$\beta = \frac{\partial P}{\partial Vol} Vol$$

- By Pressure or Density variation (in 1 dimension or direction "x"):

$$\frac{\partial P}{\partial x} = \frac{\partial \rho}{\partial x}$$

- By diameter "a".

➜ Density and Pressure, are two variables dependents:

It´s possible to explain the curvature of light by the effect of gravity (not Newtonian concept), through the concept of refraction: light curves the more the density (pressure) energy increases.

The higher the refractive index, the slower the light. ... The higher the density of the air, the more molecules there are per volume, and the more light is being obstructed. Therefore the refractive index also increases.

Energy density is a measure of the energy stored in the field per unit volume of space. ...

Let us use the similarity between the gravitational and electric fields to construct a gravitational energy density term.

➜ Light bends, due to the existence of an energy density variation; that is: the energy density, it bends the space-time and so, the paths.

The energy density, have 3 contributions:

- Electric density: J/m^3:

$$w_E = \frac{1}{2}E^2 \cdot \epsilon_0$$

Vacuum permittivity:

$$\epsilon_0 = 8.854 \cdot 10^{-12} \; As/Vm$$

field strength $E = 1kV/mm = 1 \; MV/m$

- Magnetic density: J/m³:

$$w_B = \frac{1}{2} B^2 / \mu_0$$

"B" units in "T".

Vacuum permeability:

$$\mu_0 = 4\pi \cdot 10^{-7} \; Vs/Am$$

For earth: B = 45 μ T

- Gravitational density: "g" gravity acceleration: J/m³:

$$\omega_G = \frac{g^2}{8\pi G}$$

For calculating the bend angle of light, when pass close to mass:

- "m" ("G" gravitational constant, "r" distance light to mass center, "c" light speed); in radians; applying Buckingham-PI theorem:

$$\theta = \frac{4Gm}{r c^2}$$

- By Newton theory, also is possible to calculate the same value, applying (Kinetic minus potential energy):

$$\frac{1}{2}mV^2 - \sum_{i=1}^{n}\frac{GM\,m_i}{r} \rightarrow$$

$$\rightarrow mimimum - value$$

The bend, it blends by gravity (and others forces) action, creating a big big ellipse.

Also, applying Newton theory:

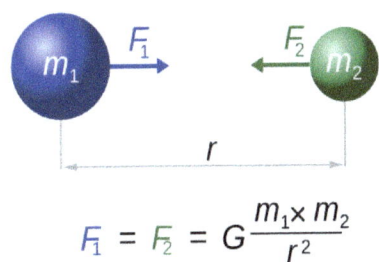

$$F_1 = F_2 = G\frac{m_1 \times m_2}{r^2}$$

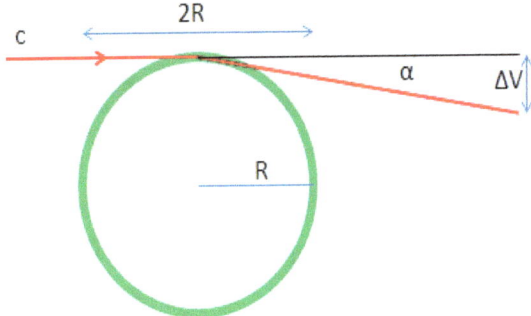

$$\Delta V = g\Delta t = \frac{Gm}{R^2}\frac{3R}{c} =$$

$$= \frac{2mG}{Rc}$$

$$\Delta V = \frac{2mG}{Rc} \rightarrow$$

$$\rightarrow \alpha = \frac{2mG}{Rc^2}$$

About the refractive index in space-time "n(R)":

$$n(R) = \frac{1}{1 - \dfrac{2mG}{Rc^2}}$$

With this value, it´s possible to calculate the curvature of light in black hole, ghost images of distances galaxies, etc....:

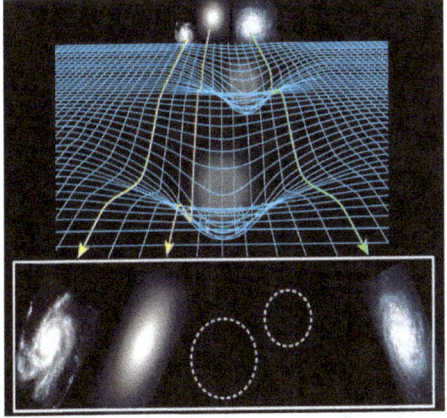

This value is half part of "real" result; so, it´s necessary some think more....

- *Temperature (T)*

Calculating the pressure for a 1 mole; then ("N_A" is Avogadro number, "M" molecular mass, "K" is a constant, "Rg" is universal constant of gasses, "u" the velocity "T" the temperature):

$$P * Vol = m\,u^2\,N_A = Rg * T$$
$$T = K * M * u^2 \propto u^2$$

From these equations:

$$P \propto \rho T$$

In this case, "u" is the average if velocity because: It is always said that the displacement of the particles or molecules of a fluid is something unpredictable: in the Brownian displacement, the particles vibrating (Temperature), produce a variation of position, and this position unpredictable, produce an evolution unpredictable.

Refractive index values are usually determined at standard temperature. A lower temperature means the liquid becomes denser and has a higher viscosity, causing light to travel slower in the medium. This results in a larger value for the refractive index due to a larger ratio.

- ***Instability***

➜ Instability is instability; the name depends of geometry created after.

In any dynamics, instability can originate; from it, the dynamics is very different; this instability is generated from a difference in viscosity, density or other parameters for example. How boring life would be without instabilities: everything would be very monotonous and uniform.

Imagine a flow of air hitting a sphere or ball; if the air has a low velocity, you will notice that the air moves "smoothly" around and behind the ball:

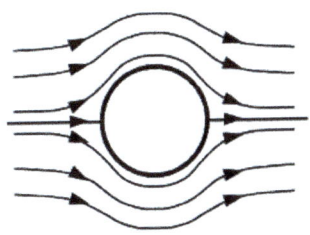

In this case, it is said that the flow is laminar; that is to say: that no eddies or generally called turbulence are appreciated; the truth is that everything would be very boring without turbulence; in fact, the Navier Stokes or Schrodinger Equations can also be applied in psychology, mass control or design of pedestrian evacuation systems in sports enclosures for example; everything would be very easy, in the absence of the turbulence phenomena.

Imagine now, that each air molecule follows another air molecule and so on; there is an endless number of molecules all following a uniform line:

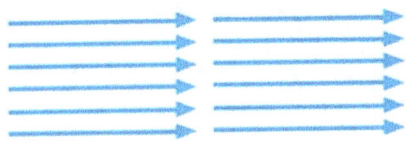

Let us imagine that suddenly, for whatever "reason", there is a single molecule that does not follow this dynamic pattern; that is to say: it deviates even very slightly from the "normal" trajectory; technically, it is said that "instability" has occurred; this instability is the beginning of turbulence; from that moment on, logically, the variations in trajectory follow one after another, as one molecule pushes another molecule to change direction, and so on and so forth. The "reason" why that first molecule changes its trajectory can be very, very varied: very subtle changes in temperature, or pressure or density, or even, most commonly, of unknown origin...:

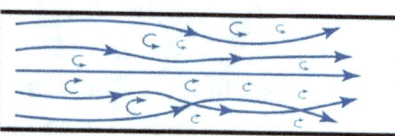

Depending on what geometry or structure is formed next, instabilities are given names.

The concept of instability is analogous to the concept of non-homogeneity.

Non-homogeneity is needed for dynamics to exist, just as instabilities are needed for particular and special geometries to be formed and transformed into dynamics. If not, all is the same.... (without changes....).

When you try to break a cake, the cake does not deform indefinitely: it breaks into cracks.

How are these cracks formed? From instabilities; the instabilities, are necessaries:

For example: if we pour sugar particles on a cake, their distribution will not be uniform; there will be concentrations or aggregations of particles spread all over the cake:

In the daily life of all of us, if there were no alterations or instabilities, not only would it be very boring but there would be no progress of any kind, no desire to advance or innovate. Everything would advance uniformly. Instabilities are essential, even in the life of the human being.

Instabilities there are a lot types (geometry created); the Kevin Helmholtz instability is typical, but not alone. For example, Karman vortex as turbulences: have a special geometry and dimensions, and is periodic.

The Kelvin-Helmholtz effect, cause a little variation between different layers of fluid; these variations change in time, producing "turbulences" with a special geometry. The variation may be caused by gravity or random displacement of molecules (Brownian displacement of molecules). On a moving surface, the boundary interface produces a fluid brake, because there are different velocities and may be densities and/or viscosities.

The origin of these geometries is simple: start from a little perturbation; dynamic process:

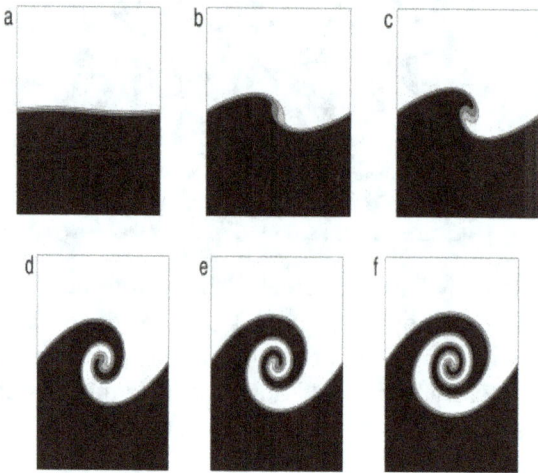

Also, it´s possible that from these specials geometries create the galaxies arms….

So if that disturbance does not exist, it is necessary that the 2 bands circulate fully parallel. But it is more likely that not circulate parallel. In this way, disturbance occurs.

If two fluids bands are different density or/and viscosities and velocities will be a disturbance: the fluid with more velocity, tend to fill the low pressure in a fluid with less velocity.

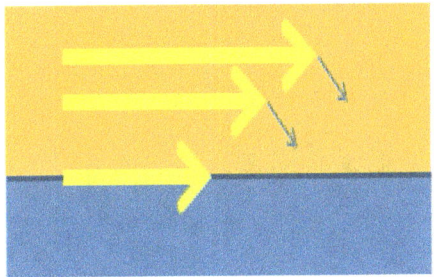

This low pressure zone is originated by:

- Gap time of reaction between molecules; that is: different viscosity.
- Different velocity (simple, but real).
- Density different.
- Gravity, if exist (or centrifugal force, etc).
- Rayleigh-Taylor:

This last effect, occur when a fluid with more density, interaction with other fluid (by gravity or inertia force):

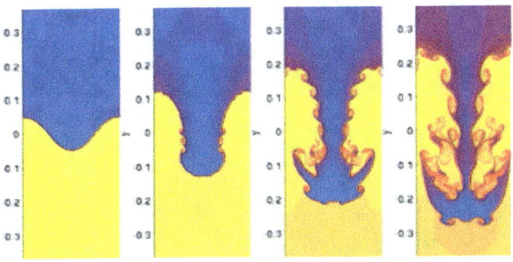

This instability, which gives rise to Rayleigh geometry, is very similar to the following event:

A glass full of water; we tilt it and when a little water leaves the glass (falls out), almost all the water in the glass also falls out; this is what happens with this instability: it drags a large amount of fluid downwards (with gravity and cold fluid) or upwards (with gravity and hot fluid):

Considering the origin of turbulence in terms of small initial disturbances, one case where we can see and observe the creation of turbulences is the curtains of most rural houses. It all has seen these curtains which are placed on the door to prevent the entry of mosquitoes. If it's windy, we will see that the curtain starts to ripple. Originally, the curtain doesn't move, but with a slight alteration (instability), the wave starts:

Perturbation of Kelvin-Helmholtz also in other´s structures bigger, as Orion Nebula or even in the Sun surface:

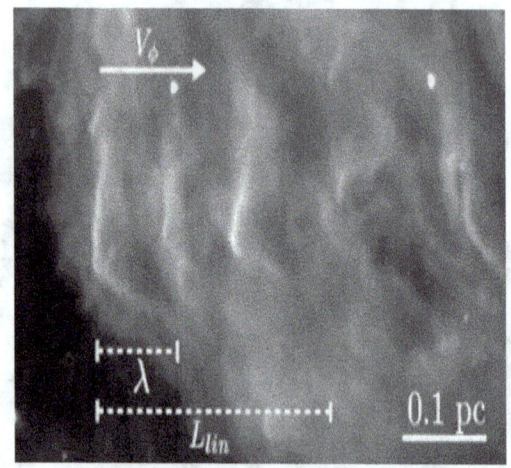

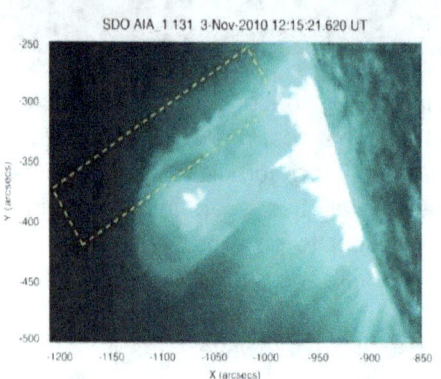

These perturbation of Kelvin-Helmholtz, are very spectacular; we can think about this disturbance, as a brake wave:

It´s possible to see these specials geometries in stones (flow sand, etc); the origin, is the same: first little disturbance, and after density-viscosity differences or others; Kelvin-Helmholtz disturbances in stones or mountains:

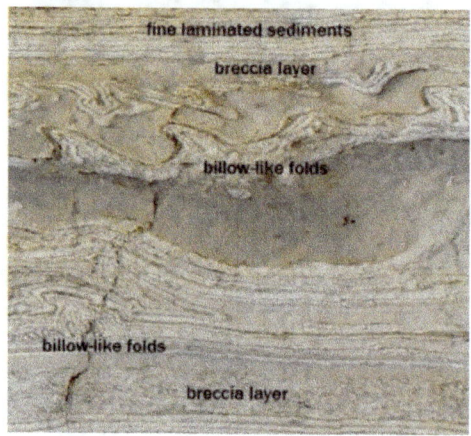

These beautiful folds, which are repeated in different contexts, are analogous to the waves of the sea; in the sea, there are small waves, there are also bigger waves even so big that they break, like Kelvin Helmholtz's "vortices":

About the Von Karman Instability:

When an air flow circulates around any geometry, it evolves around it, causing instabilities; these instabilities, as we have already seen, form turbulences; these turbulences are of practically infinite types and forms; most of them are not periodic, that is to say: they are not repeated in time and space, but there are some that are; this is the case of the so-called Karman Vortices; they only form under very specific conditions of air speed together with certain measurements of the object.

For Karman Vortices to form, the so-called Strouhal number ("S") must be around 0.2:

$$S = \frac{fd}{u}$$

"f" being the frequency of the vortices, "d" the diameter or characteristic dimension of the object and "u" the air velocity.

What is the origin or process of creation?

Scientists have the mania of wanting to know the origin of any phenomenon; and what better in this case, than to know what happens just at the beginning of the event; let's suppose that a flow of air hits a sphere; when surrounding it, both in zone "A" and in zone "B" a low pressure (or low density) tends to be produced; but it is impossible for them to occur at the same moment; so always, and by means of an "Instability", in one zone the pressure drop occurs first (for example in "A"); the airflow will tend to accumulate in "A" and then in "B" since a pressure drop will also occur there, and so on:

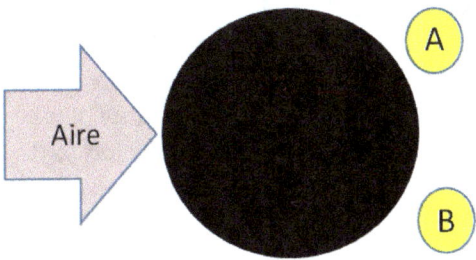

They are so called because they were first studied by Theodore Von Karman; when he analyzed them, he said: "I found that only the antisymmetric arrangement could be stable, and only for a certain proportion of the distance between the rows and the distance between two consecutive vortices in each row", i.e. the pressure drop in "A" and "B" is not simultaneous.

Behind an island, for example, a series of periodic vortices are formed with an anti-symmetric arrangement, as Von Karman said; note that the vortices "A" and "B" have opposite directions of rotation, because the viscosity makes them drag each other, like gears in an engine:

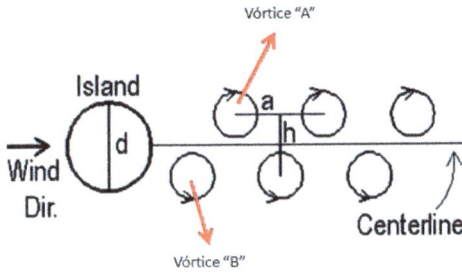

The advantage of these vortices is that they always comply with a series of geometric measurements and periodicities, i.e. they follow a pattern; this allows, among other things, not only to analyze and compare them, but also to design, for example, anemometers (apparatus for calculating air velocity); i.e. if we know the periodicity as a function of velocity, if we know the periodicity of a flow, we can know its velocity: if we know the periodicity as a function of the velocity, if we know the periodicity of a flow, we can know its velocity; for this, the anemometer in question must have a kind of cylinder inside it, which is crossed by the air flow, measuring the periodicity:

$$\frac{fd}{u} = 0.198 \left(1 - \frac{19.7}{\text{Re}} \right)$$

"f" is the frequency or periodicity (cycles) per unit of time, "d" diameter of the cylinder, "u" flow velocity, "Re" is called Reynolds number of the phenomenon.

In civil engineering for example, these periodicities are not good; the reason is simple: suppose a chimney on which Karman Vortices form downstream; each vortex formed downstream produces a force on the chimney and the alternation of turbulence at one place and another causes the chimney to vibrate; if this vibration coincides with the so-called resonance frequency of the chimney, it can bring it down; no more, no less: bring it down. To prevent this, the chimney is surrounded from top to bottom with a spiral ribbon:

The Tacoma Narrows Bridge in 1940 fell down because of these Vortexes; it is no joke:

- ***Viscosity (μ)***

The Viscosity seems a friction force in order to stop the dynamic or particle movement.

For example, whether "A" is an event (oil price), represented by its phase diagram, depends on 2 variables: productivity (number of barrels "Nb"), "Kw" (Kilowatts) per day produced by alternative energies; Time "t" is always present:

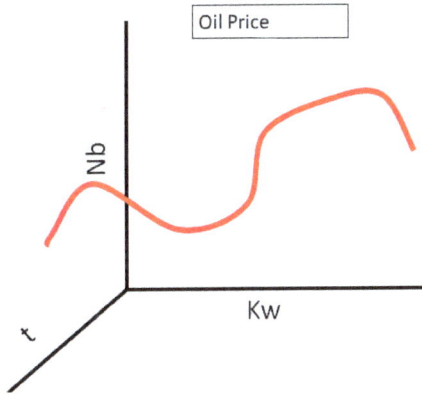

When starting a car when the traffic lights turn green, it does so after some time after the car that precedes it moves (delay time or "T_d").

It also happens when the price of oil changes due to the index variation of the New York Stock Exchange-Market; it does not do it immediately. "Viscosity= $\mu = 1/T_d$".

If $\mu = \infty$, if and only if, ρ = Constant.

It can see the same delay or gap time, in a typical prey-predator numeric model, between input and output (excitation and response - pick to pick) or in oil price against politic decision:

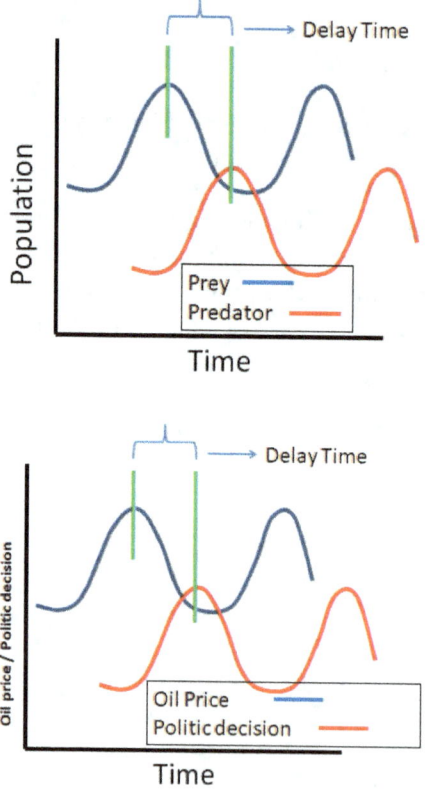

1/Viscosity, as reaction time or gap time, brings together the reaction times of all the forces involved in the displacement of a particle: the force or gravitational field, induces a reaction time, the same as the magnetic field, pressure and others; the "final" viscosity, is the reaction time of a particle, before all the force fields that work or act on the particle (adding all delay times).

It defines Sound vibration or Sound Wave: in a fluid formed by particles or mutually dependent events, a Sound Wave is defined as the evolution in a direction "V" of a variation between the particles.

Viscosity = 1 / delay time between molecules in a fluid, in order to transmit the sound in a direction "V"; it is a way to classify different fluids or events.

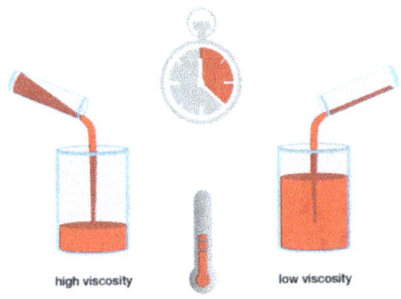

high viscosity   low viscosity

→ Particular case:

Calculate now, the reaction time between 2 particles of a fluid (in this case, fluid is a fluid "traditional"), for transmit a sound wave. Coordinates of this event: "C" is the speed of sound (wave shock) in a fluid, "R" is the fluid constant, "x" is the average displacement of particles (as a Brownian movement), "t" is time and "N_m" the number particles in 1 lineal meter, "N_A" is Avogadro number:

$$T_d = \frac{1/C}{N_m} = \frac{1}{C\sqrt[3]{\dfrac{P}{RT}}\, N_A} =$$

$$= \frac{1}{C\sqrt[3]{\rho N_A}}$$

Einstein viscosity value is:

$$\mu_E = \frac{RT}{N_A} \frac{1}{6\pi Dr} / x^2 \propto D * t$$

"D" is Diffusivity and "r" radio molecules or particles. So:

$$T_d = \sqrt[3]{\frac{\mu_E 6\pi Dr}{P C^3}}$$

This "gap time" is a vector, because in every direction, will be a 1/viscosity.

Is possible so, in this moment, to do a fluids classification against "T_d". For that, is necessary calculate all with the same pressure and temperature. The sound speed "C", for every fluid, depends of variation of pressure, against density; that is:

$$C \propto \sqrt{\frac{\partial P}{\partial \rho}}$$

This expression is equivalent to: the sound speed, depend of temperature "T". That is very important:

$$C \propto T$$

From another point of view, it has a particles group and between them, there is a spring between particles (or full fluid volume) with a constant "K"; from Hookes law, it is ("x" displacement, "u" velocity", "t" time):

$$F = Kx = m\frac{u}{t}$$

$$K = \frac{m\,C^2}{N_A\mu_G} \propto \frac{m}{\mu_G}$$

This "delay time" or "delay phase" (between input and output signal), can produce Lissajous curves: for example:

In a flapping wing case, show a position against lift generate by wing, for a one frequency; show "input" and "output" and delay time between them ([30] Timoteo Briet Blanes); other sample about: incubation time for coronavirus or Covid-19:

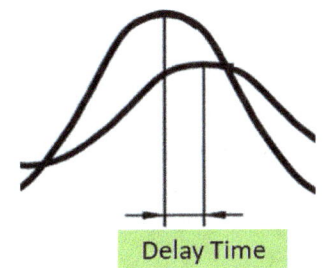

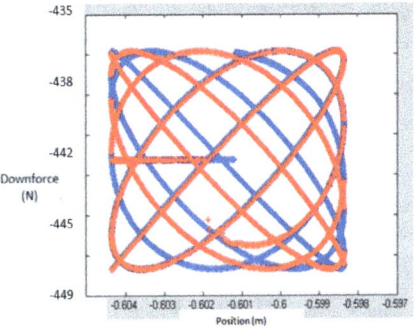

→ This Viscosity, seem the "Friction" (force), which in my opinion, is the mother of all properties.

This Viscosity, is a Vector, because can be different in each direction.

What is the Diffusivity "D" as a fluid property? Is the tendency to fade. If it have a spherical particles group ("r" radio particle, "cte" is a constant). When the viscosity and radio particle is greater, the diffusivity is less, if "T" (temperature) is greater the diffusivity also; that is:

$$D = cte * \frac{T}{\mu r}$$

Comparing this, with the Einstein relation for diffusivity ("K_B" is a Boltzmann constant) (very similar):

$$D = \frac{K_B T}{6\pi\mu r}$$

The viscosity is a function of Density; that is: the Viscosity, depends of Density by a function "f":

$$\mu = f(\rho)$$

This "gap time" as a Viscosity, is a Viscosity total, as a sum of all types of viscosities.

As for Viscosity, there are more types of Viscosities, not only the one directly related to a fluid, which also influences the dynamics of a particle:

- Gravitational Viscosity.
- Electromagnetic Viscosity.

For Gravitational Viscosity, an experiment can be carried out to calculate it:

A spaceship crosses the space between two planets of equal mass, through the central part. This is so that both planets produce the same amount of gravitational density.

In the case of planets of different masses, the spacecraft must pass through the area of equal gravitational attraction between the planets.

As it passes, the ship will have a reduction in speed; in this way, it will be possible to calculate the Gravitational Viscosity: it´s necessary so, calculate the variation of velocity, against the gravity acceleration and velocity:

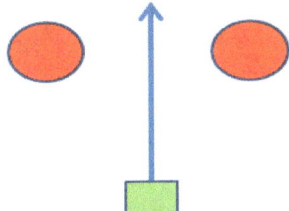

For Electromagnetic Viscosity: the same test the Gravitational Viscosity, but in electromagnetic field.

For example, if it drop a ball magnetic in a tube aluminium, the speed of ball, is reducing; that is a cause of magnetic viscosity:

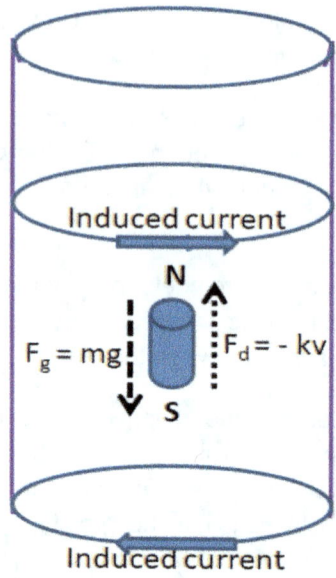

Even is possible to know the viscosity, density and viscosity/density of dark matter (plus baryonic matter) in a galaxy, in order to create a "real" velocity rotation curve ([10] Timoteo Briet Blanes):

The velocity radial in a galaxy is different if it supposes the rotation with the Kepler laws; the velocity "real" is greater.

Solid line is a velocity by Kepler rules, and star line, velocity "real":

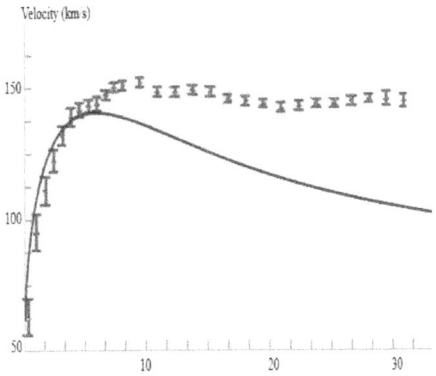

In order to create a model for galaxy matter, which its velocity profile is the correct, is possible change galaxy density and/or viscosity.

- Density (that is: considering as a only fluid, a galaxy):

The existence of dark matter becomes necessary:

The velocity "V" in a point with "r" distance to galaxy center is ("G" gravitational constant, "m" mass of particle-point):

$$V^2 = \frac{Gm}{r}$$

Sup. "Spheric" shape:

$$m = \rho * vol = \rho \frac{4}{3} \pi r^3$$

$$\rho(V, r) = \frac{3 V^2}{4 \pi G r^2}$$

Sup. "t_{orb}", the orbital time; for a "r" radio:

$$t_{orb}(r) = \frac{2\pi r}{V}$$

$$\rho(t_{orb}) = \frac{3\pi}{G t_{orb}^{2}}$$

The variation of velocity "real" with the Kepler hypothesis, is suppose that m/r, change in a special form (constant "cte"); so: ("cte" is more or less constant: is possible calculate a Density in every point, or full galaxy):

$$\frac{m}{r} = cte$$

$$2V\partial V = G(r\partial m - m\partial r)$$

Adding this "new" matter (against "r") to galaxy is possible obtains the good velocity profile; that is: increasing the density. The density value for that is approximately $1.38 * 10^{-17}$ Kg/m^3. In our solar system, there is Dark Matter, yes.

The amount in our solar system (we suppose a sphere of solar system diameter) is more or less the mass of Pluto ($1.5*10^{22}$ Kg):

Problem: We not know the size of particle dark matter: May be Pluto size or less....

- Viscosity (that is: considering as a only fluid, a galaxy):

In a galaxy, "a" star radio, "r" distance star to galaxy center, "m" the galaxy mass (inner part) and "μ" the viscosity: the velocity "V" is proportional to "a" and "r" and proportional to "1/m"; "μ" is the proportional constant (more or less: it's possible calculate a Viscosity in every point, or full galaxy) or factor:

$$V = \mu_1 \frac{ar}{m}$$

$$\mu_1 = \mu 6\pi$$

Is possible so, to know the viscosity, for having a velocity.

- Density and viscosity (working together):

Combining and substituting the mass for density*Volume (Vol), it is possible change density and viscosity, in order to have the profile velocity "real" ("Vol" is the galaxy volume inner part, "K" constant more or less):

It's possible calculate a Density and Viscosity in every point or full galaxy): (the Viscosity is more important than Density):

$$\sqrt{\frac{Gm}{r}} = 6\pi \frac{a}{m} \mu r$$

$$m = \rho * Vol$$

$$\rightarrow \frac{\rho^{1/2}}{\mu} \approx K$$

In the 3 cases, is possible create the real profile. Solid line is a velocity by Kepler rules, and star line, velocity "real":

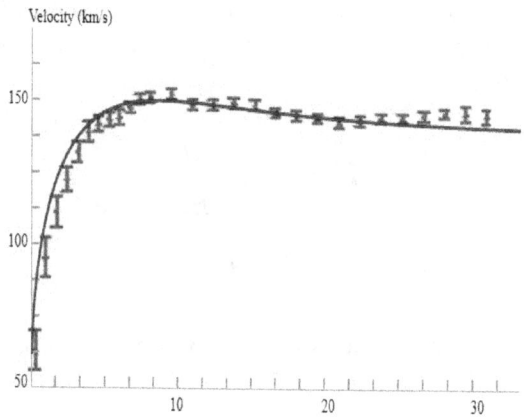

In this case of viscosity, is possible "ADD" the viscosity, to gravity and others, in order to simulate the creation and evolution of a galaxy. In fact, in practically every galaxy, it can find a dark matter with viscosity as a friction force ([43] Mahmood Roshan, Bahram Mashhoon).

In the early universe, there was less dark matter than today.

From theories MOND or Verlinde types, the dark matter is "created" by baryonic matter: Speed rotation in galaxies in early universe:

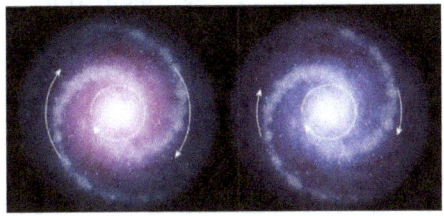

The existence of dark matter is "not necessary" for Mond. Perhaps, the expression of the gravitational force is erroneous at great distances, and the transmission of these forces, is appreciable far from the source.

For this, in the expression of Newton's gravitational force, we can include a factor that depends on the distance, but that, like the Lorentz equations, is maintained along an infinite distance, but at a short distance, as it is the case of our solar system, the "new" force and the Newton, are practically the same:

Only depend of the acceleration.

In fact, in our solar system, the acceleration is practically zero. It supposes the existence of Dark Matter.

There are others Theories in order to substituting the Dark Matter:

- Existence of lot Neutrinos.
- Machos.
- Wimp.

-   Etc…

None more comments about these "news" theories…. (As Richard Feynman say: "if there are not evidences in reality, are wrongs ¡¡¡¡").

Finally, the viscosity depend of temperature; there are a lot expressions for that; in liquids, the viscosity is lesser if the temperature is greater (Andrade relation); if gasses, the opposite (Sutherland relation) (A, B, a, b are constants):

$$\mu(T) = A\,e^{B/T}$$

$$\mu(T) = \frac{a\,T^{1/2}}{1 + b/T}$$

-   ***Similarity number "S"***

In order to be able to compare phenomena with each other or simply to know limit or transition values between different dynamic states, a value is needed. This value is denoted as "S_n".

For example, in evacuation systems pedestrians, is possible to define other phenomenon number; this is a case particular: pedestrian group in a room with exit of "A" dimensions; red arrow is the people direction evacuation:

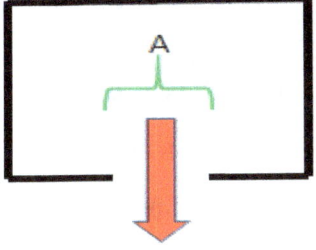

"V" velocity of pedestrian, "A" length of door, "ρ" density of pedestrian group, "T_d" delay time or reaction time between pedestrians; this value, it can apply to fluid in a duct:

$$S_n = \frac{\rho V A}{\mu}$$

This value, not have dimensions; must to be as that: density → number people per square meter, Velocity → meter per second, "A" → meters, "μ" → 1/time.

Is necessary now, calculate the value for the Similarity number, from which there is a accumulation of peoples dangerous in exit (may be change from laminar to turbulent).

For calculating the gap time: given a group of people, we push or move one of them; the time it takes (on average) for the people around it, will be the gap time "T_d"; this test, in the same density conditions, as the problem to solve. In short, it is a problem of calculating the speed of transmission of a pressure wave.

Another case where the use of this dimensionless value can be observed: it is assumed that the appearance of an event in a given time window is analyzed; the resulting value is dimensionless ("A" in time).

- **Tension and Expansion (TE)**

The tension "TE", is the force that exerts a fluid to not expand (more separation between particles); can be understood as the force that the molecules perform, not to expand. As always, the tension and the expansion are calculated in a direction "V".

In the case of matter (fluid) in space, the same concept can be understood as gravity, since the greater the gravity, the greater the force to be expanded.

There is an expression very important (general expression), in order to have a relation between surface tension, compressibility and density; that is ("TE" is a tension, "Z" compressibility, "ρ" density, "K" a constant):

$$TE\left(\frac{Z}{\rho}\right)^{1/2} = K$$

In fact, the tension is a drag force, as a viscosity, to movement.

Can be defining the tension, by ("T" temperature, "P" pressure, "Vol" volume):

$$\frac{\partial P}{\partial T} \ or \ \frac{\partial Vol}{\partial P}$$

There is other parameter very important, with very relation with the tension; it's the Expansion force "α" of fluid ("Z" compressibility, "Vol" volume, "T" temperature, "ρ" density,"$(M)_n$" variable "M" with "n" constant:

$$\left(\frac{\partial P}{\partial T}\right)_{Vol} = \frac{\alpha}{Z}$$

$$\alpha = \frac{1}{Vol}\left(\frac{\partial Vol}{\partial T}\right)_{P} =$$

$$= -\frac{1}{\rho}\left(\frac{\partial \rho}{\partial T}\right)_{P}$$

$$Z = \frac{1}{Vol}\left(\frac{\partial Vol}{\partial P}\right)_{T}$$

ARTICLE 3

Matter Aggregation

Any aggregation or grouping of matter begins with an instability in the dynamics of a group of particles; if there were no instabilities, any dynamics would be linear or laminar (very boring dynamics); instabilities produce or originate groupings of matter; the reasons behind these instabilities are of various kinds.

- ***Aggregation by Viscosity***

It can see the accumulation of dust and lint at home, in a dispersion of tree leaf by the wind, in an accumulation of drop water in a flat plate, in clouds or plastics in sea:

The friction between particles is the responsible of these accumulations: that is: the Viscosity.

Even, it´s possible to see the self-classification or self-ordination of different seeds types, by viscosity:

The same occur in "flock's ice" in sea, petroleum-oil filaments in leaks or water in cascades:

A case very interesting for me, is the Article, write by: ([32] Tsuyoshi Inoue). This article-simulation was performed by Tsuyoshi Inoue, with the ATERUI supercomputer operated by the @prcnaoj_en.

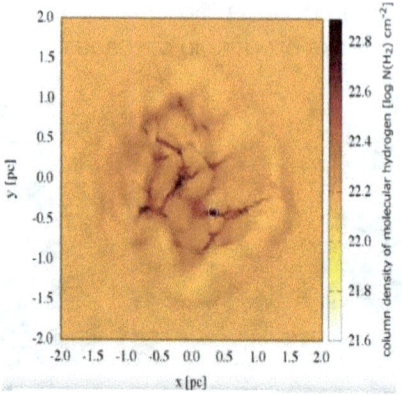

It can be observed that the seeds (yellow) accumulate in groups. This is due to the friction between them (viscosity). If these seeds are in a garden with grass, the accumulation in groups does may be not occur: the distribution may be is uniform.

This is because the grass has a higher friction (or viscosity) or in general, different friction; this fact, have consequences (it will see these):

So, the environmental it´s very important also.

The environmental have the conditions ideals for accumulations of people; may be that this accumulation is due to the same feelings (to be in beach....); or accumulations of seeds in step street:

A lightning is a trajectory that could be considered as a conditioned Brownian displacement. In the following image, it can see the influence of the rainwater (environmental) cascade on the path of the lightning, which follows the path of the water:

The same in the next image: the environmental, produce the aggregation (shadow / need of shadow):

It's been barely 15 minutes since I got up from walking my dogs and, being in a square, I started laughing (and also now....) when I found some piles of seeds on the grass:

When I saw them, I wondered if the 4 accumulations I saw and their relative position had anything to do with:

- The size of the seeds.
- The friction between them.
- The speed of the wind.
- The friction with the grass.
- The size of the clumps.
- Etc...

No. It has nothing to do with that.

At first, the seeds are rolling on the grass, and at a certain moment, for a reason that it's difficult to know...., one of them stops and drags more and more seeds to adhere to it, forming an accumulation or group.

Even well-defined accumulations of leaves can be seen along the street. The leaves themselves define the outline of the accumulation:

A typical case of the existence of main flow lines, currents or large turbulences, depending on the context where they are found or as a function of main flows is this:

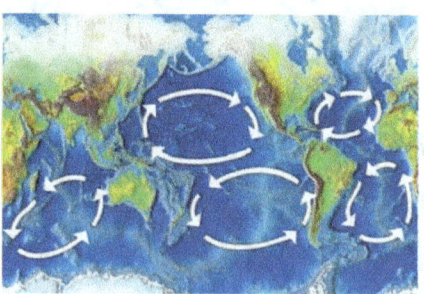

Upper and lower turbulence, are created from a main flow that runs from right to left; the smaller the scale of analysis, the more and more through a CFD simulation, it is very easy to observe the creation of currents or turbulence, from others:

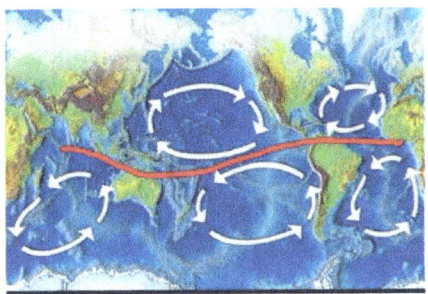

This is a case of the creation of large turbulences, but the same happens with small turbulences, originated by a larger one:

It's also possible to create turbulence, the upper one for example, from the lower one.

The particles with the same or similar viscosity tend to join (that occur also in humans….).

In the case of flock´s birds, the friction force or Viscosity, work as feelings:

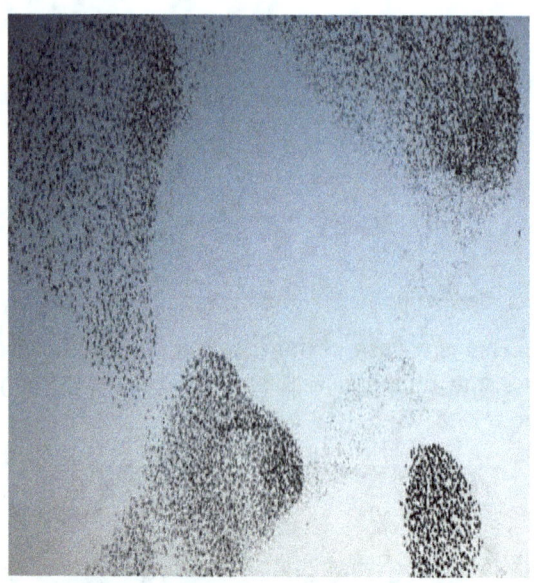

The governing equations or relations, between birds, in order to create the flocks:

- There is a "bird boss", which do the way.
- There is a cohesion or density or pressure between birds.
- The direction of displacement is "@Pressure/density (acceleration).
- Other law….

In the other hand, for example, in a discotheque (or beach before), the people is there because they to have a good time, and the place, is the right one for it. Is possible show the next images, about interfaces between fluids of different viscosity (sea water), creating filaments, accumulations or aggregations; this densities and viscosities different, may be produced also by different temperature, salinity, etc….:

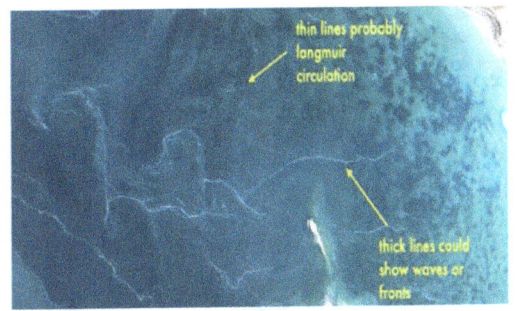

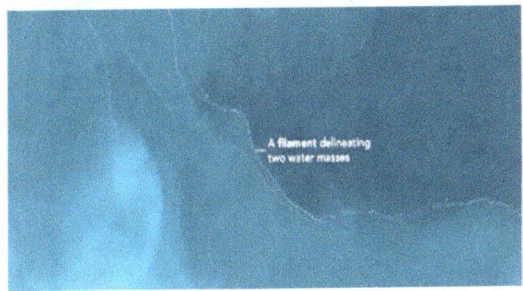

Obviously, this aggregation by Viscosity, work in the galaxies formation and others events; but always work by friction.

About this aggregation origin, I want to simulate some examples in CFD, in order to extract, may be..., some conclusions; these conclusion, are thought to priori....:

→ 1 fluid in movement (any movement) in a box; analyze aggregations or density variations by viscosity, density, combinations, etc.....

→ 2 fluids in movement (any movement) in a box; analyze the aggregation or density variations by viscosity, density, combinations, etc....

→ ....

- ***Aggregation by others forces***

It´s possible to do a group of "all forces in universe":

- Gravity (attraction):

Gravitational Force

$$F = G\frac{m_1m_2}{r^2}$$

F = gravitational force
m_1 = mass 1
m_2 = mass 2
r = distance between centers of mass
G = 6.7 X 10^{-11} N · m²/kg²

- Electromagnetism (attraction and repulsion):

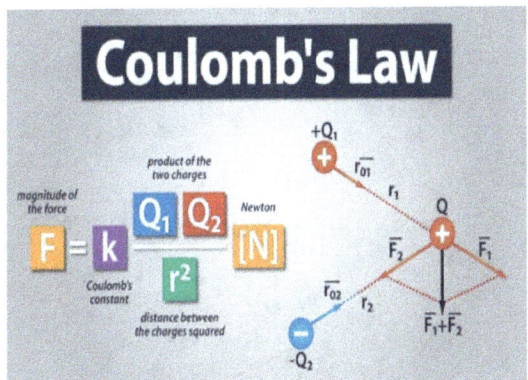

➔ The "light" also can be push, and so, affect to dynamic particles.

- Weak nuclear force.
- Strong nuclear force.

Some of these forces, work in the galaxies or in spyder web formation, etc....; basically the gravity.

- *Aggregation by Low Pressure*

A zone with low pressure (in relation other zone near), attract particles or matter; the case of a zone with high pressure, repelled or push a particle:

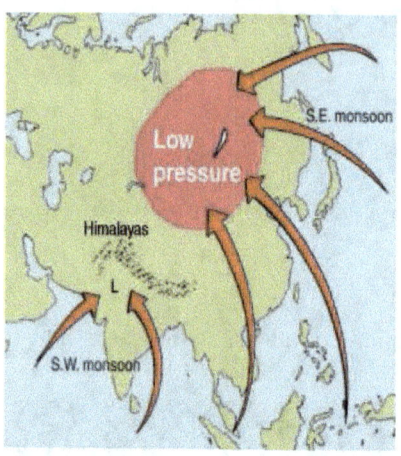

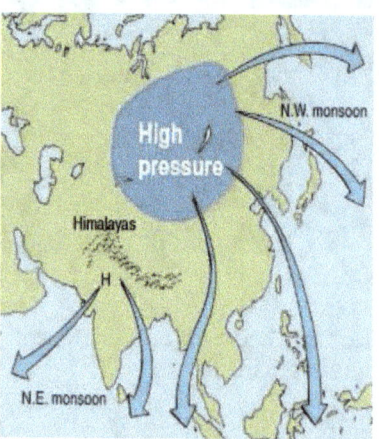

A particular case is the creation of depression tube and vortices:

When a particle move, his path is a depression zone; this zone is an attractor for any particle around; this depression tubes, create vortices around:

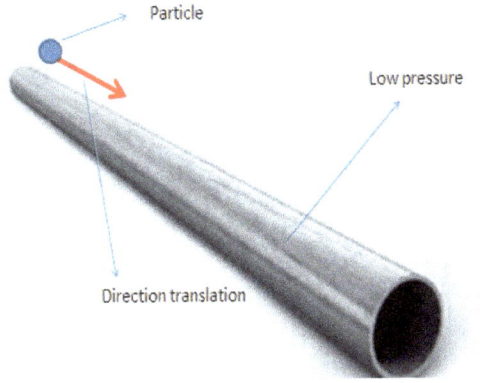

© Steve Morris

This low pressure, is present also around every particle in displacement, so, others particles are attracted.

It´s possible, create an analogy with the gravity: let a spacecraft around planet; he have a speed and the gravity force "suck" the spacecraft to planet; depending the distance, the gravity force and speed, the orbit will be different (Newton theory gravity); the trajectory will be also different depending of others forces as a viscosity or density:

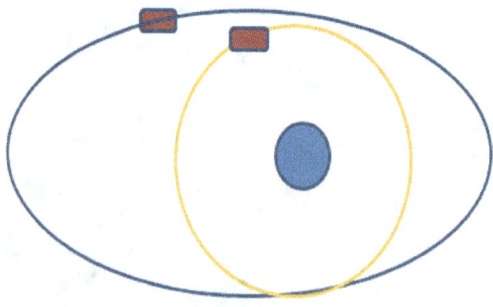

→ Some think important: the aggregation by viscosity, is similar or at least have the same origin, with the aggregation by low pressure: the particle path, produce also a low pressure zone. That is very important to know it: Article ([33] Roberto Camassa, Daniel M. Harris, Robert Hunt, Zeliha Kilic & Richard M. McLaughlin).

Analyzing these orbits around vortex center, it should be possible to know the vortex geometry, his force, his evolution, his combination, etc....

Some equations and relations about; "r" distance between air particle and vortex center, "v" speed particle, "P" pressure "attraction" center vortex, "G" constant vortex and "T" rotation period:

$$circular - orbit \rightarrow G\frac{m}{r^2} = m\frac{v^2}{r}$$

$$T = 2\pi\sqrt{\frac{r^3}{Gm}}$$

$$P = \frac{G}{r^2}$$

Let "m=1":

$$F = G\frac{m}{r^2} = \frac{G}{r^2}$$

Calculating the velocity rotation by CFD simulation and trying that the same velocity is the result of Newton theory, is possible so, know "G".

That is very very important.

If let Force by Newton = Force Centripetal, we can know the orbit behavior. That is: what is the "G" value for our case:

$$\frac{mMG}{r^2} = m\frac{v^2}{r}$$

$$\frac{MG}{r} = v^2$$

$$G = \frac{v^2 r}{M}$$

If "a" is the semi major axis (orbit), and "T" the period:

$$a = \sqrt[3]{\frac{GM\,T^{3}}{4\,\pi^{2}}} \rightarrow \frac{a}{T} = \sqrt[3]{\frac{GM}{4\,\pi^{2}}}$$

$$G = \frac{a^{3}\,4\,\pi^{2}}{M\,T^{2}}$$

There is a value very important for defining a vortex; that is named Vorticity; for calculating that:

1.

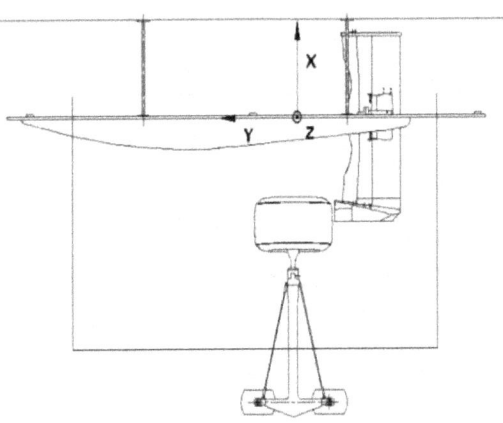

$$(\omega_{y})_{i,j} = \frac{\Gamma_{i,j}}{4\Delta X \Delta Z}$$

$$\Gamma_{i,j} = \tfrac{1}{2}\Delta X(U_{i-1,j-1} + 2U_{i,j-1} + U_{i+1,j-1})$$
$$+ \ \tfrac{1}{2}\Delta Z(W_{i+1,j-1} + 2W_{i+1,j} + W_{i+1,j+1})$$
$$- \ \tfrac{1}{2}\Delta X(U_{i+1,j+1} + 2U_{i,j+1} + U_{i-1,j+1})$$
$$- \ \tfrac{1}{2}\Delta Z(W_{i-1,j+1} + 2W_{i-1,j} + W_{i-1,j-1})$$

This last value of Circulation, is function of U and W, velocity in the two axis X and Z.

2.

Another method for calculating the vorticity, is based in displacement points in area:

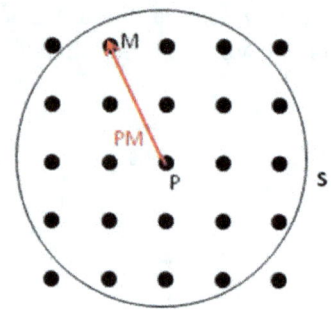

$$\Gamma_1(P) = \frac{1}{N}\sum_S \frac{(PM \wedge U_M)\cdot z}{\|PM\|\cdot\|U_M\|}$$

$$\Gamma_2(P) = \frac{1}{N}\sum_S \frac{\left(PM \wedge (U_M - \overline{U}_P)\right)\cdot z}{\|PM\|\cdot\|U_M - \overline{U}_P\|}$$

Center vortex:

Where "P" is a fixed point in the measurement domain. "S" is a two-dimensional area surrounding "P", referred to as the window size, whose size must be defined by the user. "N" is the number of data points (M) that are located within "S", and "UM" is the velocity vector at point "M".

Finally, "z" is a unit vector normal to the plane of measurement, and "Γ1" (or "Γ2") is calculated for all points in the measurement domain and then used to determine the center of the vortex. For an ideal axisymmetric vortex, the maximum value of |Γ1| is 1. For each possible center position, the "Γ1" criterion calculates the degree to which the flow rotates around this point. In this work, near the vortex core, |Γ1| was found to reach values between 0.9 and 1.

The only difference between the "Γ1" and "Γ2" criteria is that the average velocity over the window, away. This takes into account any uniform flow within the plane of rotation. With this method, when |Γ2| is greater than approximately 2/π, P is assumed to represent a point in the vortex.

Finally, another method:

3.

The mathematical model for the flow velocity in the circumference θ-direction in the Lamb-Oseen vortex is:

$$V_\theta(r, t) = \frac{\Gamma}{2\pi r}\left(1 - \exp\left(-\frac{r^2}{r_c^2(t)}\right)\right)$$

with

- r = radius,
- $r_c(t) = \sqrt{4\nu t}$ = core radius of vortex,
- ν = viscosity, and
- Γ = circulation contained in the vortex.

The radial velocity is equal to zero. The associated vorticity distribution in the vortex filament o Core Vortex direction can be found with the curt:

$$\omega_z(r,t) = \frac{\Gamma}{\pi r_c(t)^2} \exp\left(-\frac{r^2}{r_c^2(t)}\right)$$

An alternative definition is to use the peak tangential velocity of the vortex rather than the total circulation:

$$V_\theta(r) = V_{\theta\,max}\left(1+\frac{1}{2\alpha}\right)\frac{r_{max}}{r}\left[1-\exp\left(-\alpha\frac{r^2}{r^2_{max}}\right)\right]$$

Where $r_{max}(t)=((\alpha)^{0.5})r_c(t)$ is the radius at which "v_{max}" is attained, and the number $\alpha=1.25643$, see Devenport et al. The pressure field simply ensures the vortex rotates in the circumferential direction, providing the centripetal force:

$$\frac{\partial p}{\partial r} = \rho\frac{v^2}{r}$$

- **Aggregation by Tension**

The surface tension is the force that exerts a fluid to not expand. Normally this concept applies only to liquids. But this concept, applied to any fluid, can be understood as the force that the molecules perform, not to expand.

In the case of matter (fluid) in space, the same concept can be understood as gravity, since the greater the gravity, the greater the force to be expanded.

There is an expression very important (general expression), in order to have a relation between surface tension, compressibility and density; that is ("σ" is a surface tension, "K" compressibility and "ρ" density):

$$\sigma \left(\frac{K}{\rho} \right)^{1/2} = cons \tan t$$

This general relation is very important to understand the surface tension in liquids but also in gasses or matter (particles) in general.

This is the same as saying that the greater the density (if the temperature is greater, the density lower), the greater the surface tension or fluid tension:

Water

Temperature	Density	Surface Tension
0	999,8	0,0756
5	1000,0	0,0749
10	999,7	0,0742
15	999,1	0,0735
20	998,2	0,0728
25	997,0	0,0720

Compressibility and fluid tension are very close together.

It is also responsible for droplets breaking down or water accumulating in areas above a dry surface:

- ***Fingers by Viscosity, Density, Gravity (force) and Tension-Expansion***

The generation of fingers as a geometry, it´s a special dynamic or aggregation of particles.

In this aspect, it show some cases in order to understand every phenomenon; in each case, it´s possible the participation of more than 1 origin: not important that.

Basically, the origin of any fingers, it´s also a Instability; from this Instability, it create the finger.

Sample 1:

If there is a delay time between displacement molecules, lower, the viscosity is bigger; that is the case for Lava, as a flow:

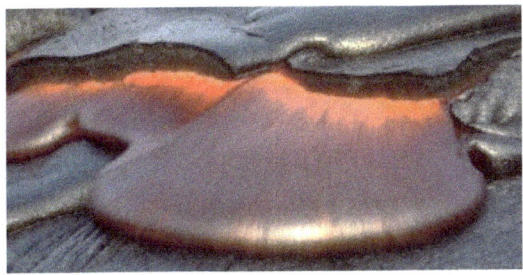

Sample 2 (analogy):

In the end of this "front shape", it can see a parabolic velocity profile....It can see this "parabolic" velocity profile, in a people group walking in street (there are boundary layer in lateral walls):

Sample 3:

Other effect very important about the Viscosity, is that:

It is possible to see "filaments" or "fingers" in the front of fluid flow as a (very similar to lava) honey for example:

Sample 4:

This is an effect between 2 fluids, with different surface tensions; this dynamic are called Marangoni Effect.

Typical sample:

Wine is a special case, as it is mainly a mixture of alcohol and water. Due to capillarity, a thin film of wine tends to rise up the walls of the glass. As this happens, both the alcohol and the water evaporate, but the former evaporates more quickly (due to its lower boiling point and higher vapour pressure), so the liquid in the wall acquires a higher surface tension than that at the bottom of the glass (remember that water has a higher surface tension than alcohol). This causes more wine to rise up the walls (to decrease the gradient of surface tension formed), so that more liquid accumulates at the top of the walls, until eventually the other force present comes into action: the weight of the accumulated liquid overcomes and causes gravity to form those tears or fingers that slide down the walls of the glass.

Sample 5:

Other example: from "Instabilities", is possible to generate fingers in beach, considering the sand as a solid:

Sample 6:

In the case of a surface with hot liquid, it is observed that instead of the liquid ascending uniformly, some "fingers" are formed:

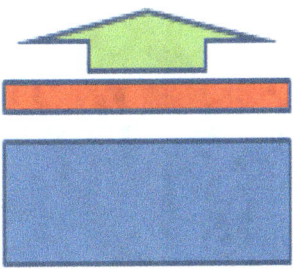

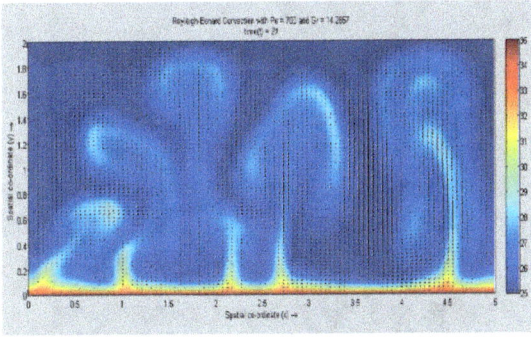

Sample 7:

Other example: two fluids, one of them in the surface; by gravity (may be any force so), the light fluid, create a "fingers" toward down:

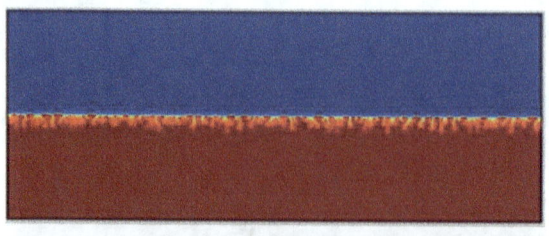

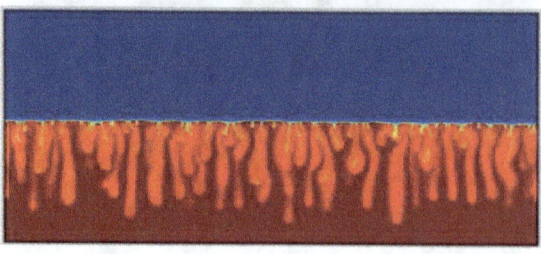

Sample 8:

The Dark Matter can be taken as a fluid of different Viscosity from the fluid that surrounds this Dark Matter and through which it moves. This difference in Viscosity and Density (and may be others factors), as it has already seen, produces a peculiar and special distribution, producing "fingers", voids, accumulations, groups, etc...

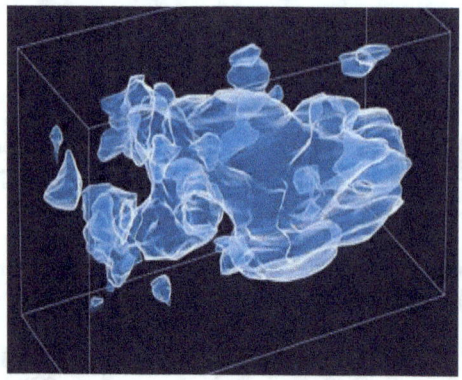

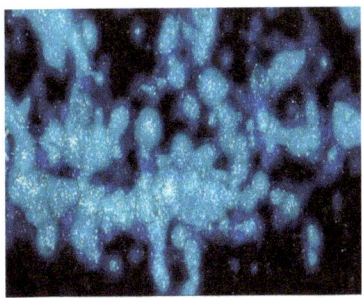

Sample 9:

The same occur in enormous blobs deep inside the Earth. It can just barely detect them using seismic imaging or Tomography, and it really don't know what they are:

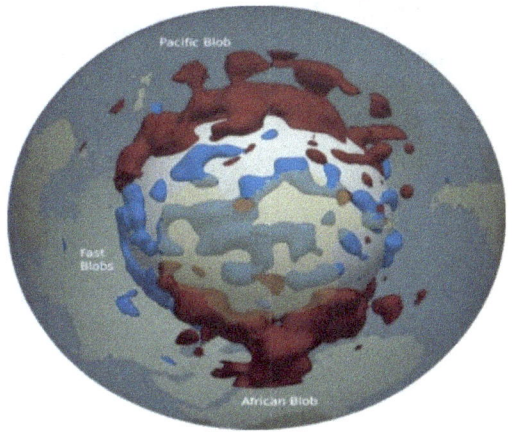

- *Supernova expansion and Nebulae Bubbles*

As a sample and similarity with the reality (fingers), it shows a supernova explosion; it´s possible to see the alterations in border: when a supernova explodes, it's always in a spherical direction. But depending on the density of the shock wave (also viscosity and more, and the environmental), even square geometries are formed;

94

it´s therefore possible to know this density variation in environmental, so that any geometry can be created; also depend of viscosity and superficial tension, between 2 fluids: environmental and star matter:

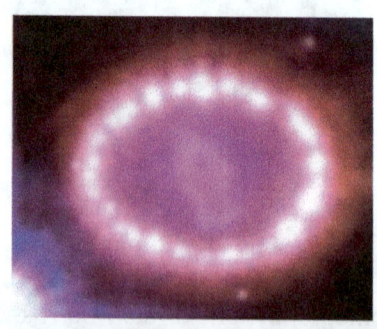

It´s possible see the same effect, in others phenomena's:

Sample 1:

It´s a sample, special; not are 2 fluids or 1 fluid on a solid....

Let's suppose that two apples collide with each other, at a speed of 90 km/h; you can see perfectly, the star structure of the skin when it explodes; the skin breaks, as it reaches its elastic limit (the surface solid of a apple, tend to expand and fill more surface....). The same applies to spherical nebulae: they are spherical until they reach a limit, from which small spheres form in the shock wave:

Sample 2:

It´s a sample, special; not are 2 fluids or 1 fluid on a solid….

In the Article ([34] Asher P. Mouat, Clay E. Wood, Justin E. Pye, and Justin C. Burton), it is possible see the same effect, in a border of fluid in expansion:

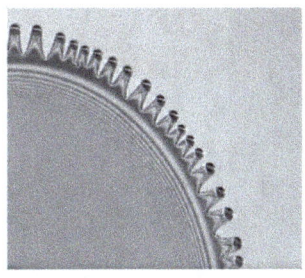

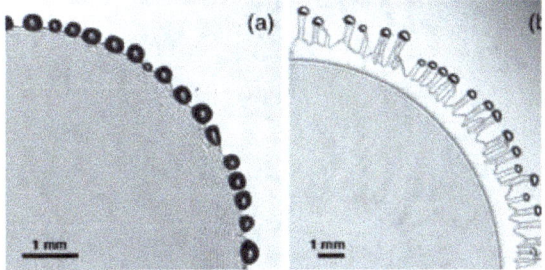

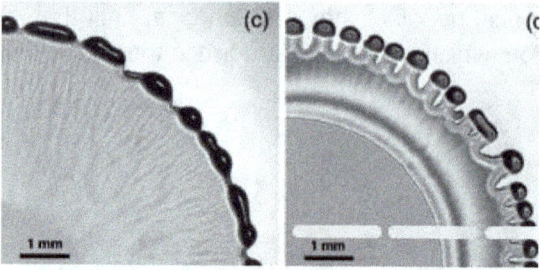

Sample 3:

     Lava explosion: the expansion is not uniform:

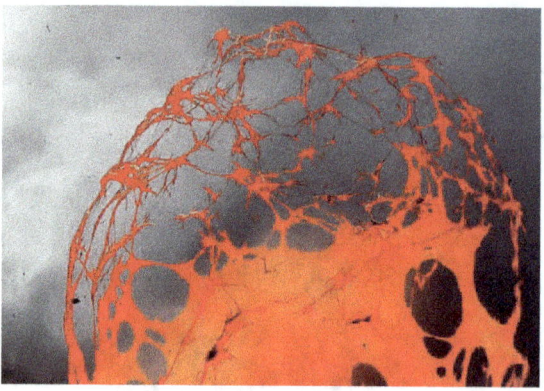

Sample 4:

     In an experiment about 2 fluid shocks: it can also see the formation of "pearls" in fluid shock:

Sample 5:

Another example can be seen on a wet solid disc, in rotation. Perhaps the logical thing would be the water covering the surface should be diffused outwards in molecules.

But the density is finite... this makes that because of the viscosity and in this case the surface tension, the water groups in point and from these points, the water escapes:

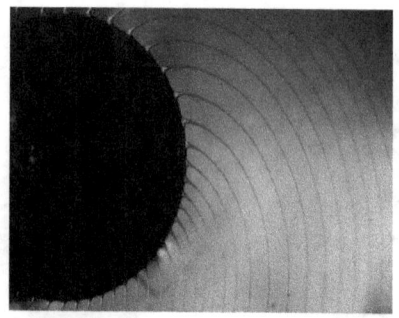

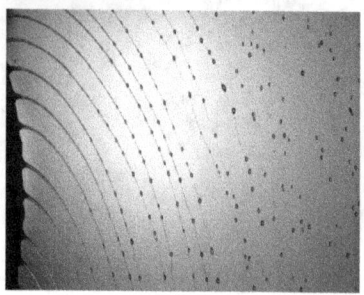

This is the same reason why "pearls" are formed in a supernova explosion so.

So, there is a limit; firm this limit, there are the formation of sub-bubbles and crack of shock wave; this "crack" limit depends of ("fluid" in expansion and environmental):

- Viscosity.
- Density.
- Surface tension.
- Size.
- Speed of expansion.
- Others like temperature, pressure, etc..

The particles of a fluid with a higher viscosity will respond more quickly to the changes in the surrounding particles; therefore, when there are small fluctuations or instabilities, they drag particles.

But, there are, for example, Bubbles nebulas. Why it is possible?

As said before, a fluid in the form of a sheet, when expanded, groups together.

But we know, for example, bubble nebulae or soap bubbles: these keep the surface, until a certain moment. This fact has limitations or conditions.

In this case, the crack "limit", is not present "yet":

If some instant, the crack is overcome:

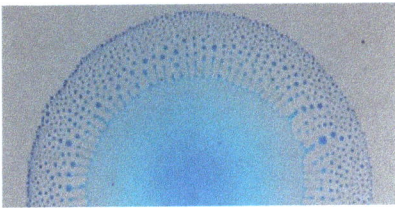

An analogy can be observed with a social fashion: fashion is advancing in Society, also advancing in Time, and there comes a moment, due to various reasons, when fashion, is divided into various fashions or trends.

ARTICLE 4

Galaxies dynamic and high scale geometry universe

- ***Galaxies: history vital***

The formation and evolution of a galaxy and galaxies can be simulated, assuming that it has working with a fluid with certain density, viscosity, etc.... conditions.

Obviously, the dynamics of galaxies is also determined by forces not related to fluid dynamics, such as gravity, electromagnetism, etc. What we are trying to do here is to give a view of the dynamics of galaxies, assuming that they behave like a fluid, inside other fluids.

Origin:

When a particle move, his path is a depression path or zone; all particles behind or around her, rotate around the depression tube, producing vortices:

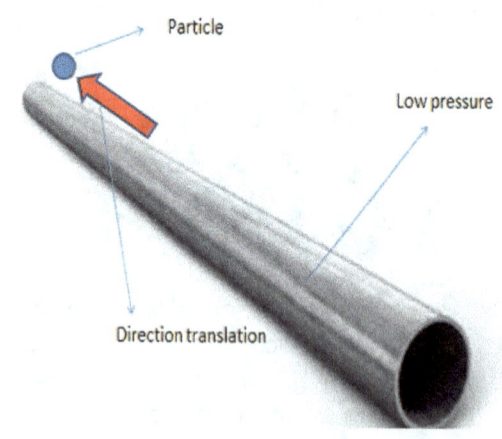

These "paths" of low pressure or density, can help matter particles, to aggregate each and other, producing, in the future, stars, black holes, etc. That is: the galaxy seed. Also, these paths allow and help other´s galaxies behind to follow the first galaxy (galaxies group). It's been barely 15 minutes since I got up from walking my dogs and, being in a square, I started laughing (and also now....) when I found some piles of seeds on the grass:

When I saw them, I wondered if the 4 accumulations I saw and their relative position had anything to do with:

- The size of the seeds.
- The friction between them.
- The speed of the wind.
- The friction with the grass.

- The size of the clumps.
- Etc...

No. It has nothing to do with that.

At first, the seeds are rolling on the grass, and at a certain moment, for a reason that it´s difficult to know...., one of them stops and drags more and more seeds to adhere to it, forming an accumulation or group.

Creation and rotation:

May be, in the center of every galaxy, exist a black hole or not.

A Galaxy rotates, because the black hole or condensed matter in its center is most likely to rotate. This turn "drags" more and more matter, also making it rotate: for example, in the case of a sump, the water starts to rotate because it is most likely (it is very difficult not to rotate....). This rotation makes more and more water turn.

The same explanation it has for a water sink: the rotation direction (without Coriolis and more), is random; there are a rotational moment, add for every particle (star) which is pull.

Black holes may or may not form in the interior of a galaxy; a black hole may be the origin of a galaxy as explained above, but its existence is not necessary; there can be galaxies without a very massive black hole in their interior just fine.

The stars, or the galaxy in general, do not necessarily revolve around the massive black hole that may exist inside it; the gravity of that black hole, if it exists, is very little in relation to the gravity of the whole galaxy; its gravity "only" affects the stars around it; the galaxy revolves around a center of mass; in fact, the massive black hole, if it exists, may not necessarily be located around the center of the galaxy.

The aggregation matter origin:

-   Created by Globular aggrupation (may be from an only one star….).
-   Dust cloud (also by dark matter aggregation ??¿¿).
-   Depression tube.
-   Friction or viscosity.
-   Black hole. This black hole, can be formed before (as a seed) or after galaxy.

Can be the globular cluster, the origin of every galaxy?: a globular cluster, have rotation (also, is the most likely....). Rotation velocity stars in globular cluster, with the rotation axis:

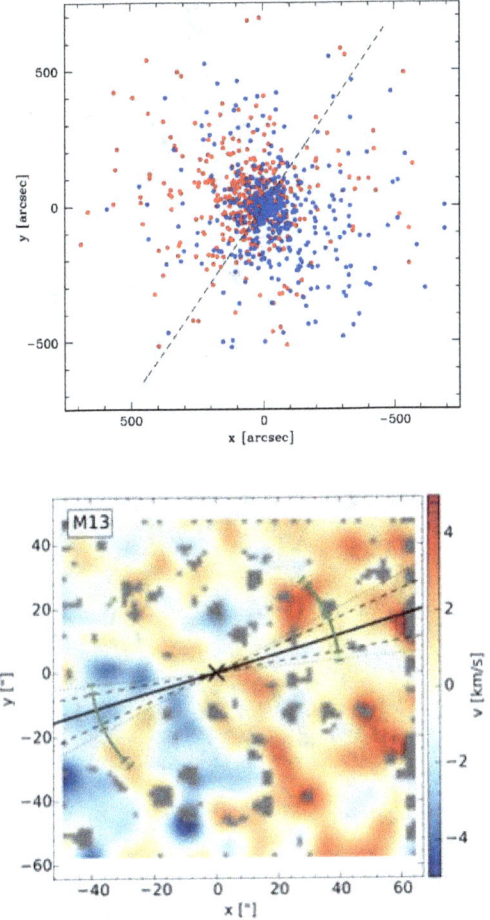

The old galaxies, are very irregulars: 13-8 billion years old this galaxy-s light comes to us from when the Univers; these galaxies are very young and not big time for development:

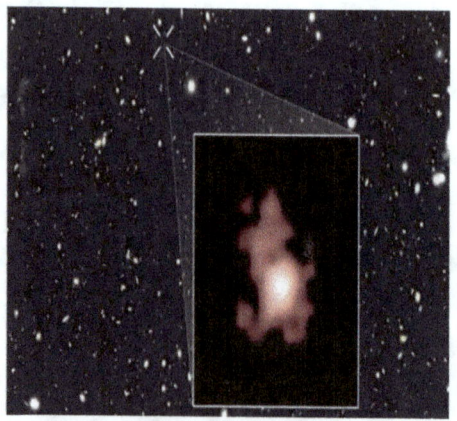

More very old galaxies:

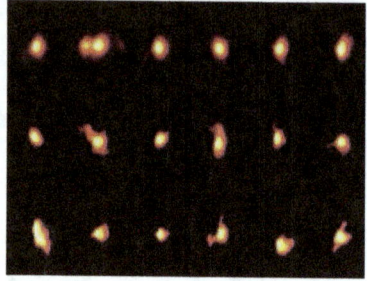

Structure formation: arms and more:

First of all, and obviously, it´s possible appreciate the similarity with any real galaxy (geometry pattern – typical in mathematical or physical models). Typical galaxy in spiral:

Is possible so, apply fluids theory to galaxy formation, evolution and interaction? May be....

Other explanation, is more accurate and "real", but complementary to these before: around a galaxy, there are a lot matter in different forms (in fact, the galaxies, swim in Dark Matter: the quantity of dark matter, is much bigger....): traditional or visible, dark matter, may be etc.... that is: the density and also the viscosity, around and into a galaxy, is big.

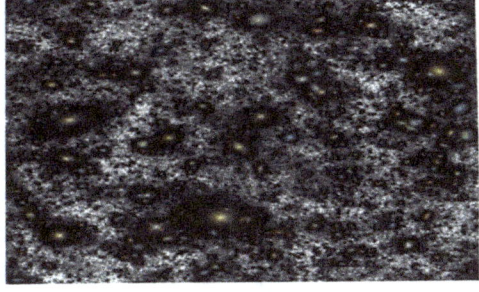

Imagine now, that it has a cloth and a cylinder placed in the central part which rotates, creating a special pattern (A rotating fluids set with different viscosities and densities, can generate arms, as a density contours). Zones of creation of galaxy arms; the same phenomenon in whirl water. Also is possible generate these "arms" in chocolate or coffee rotating.:

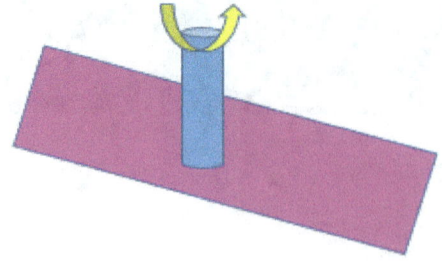

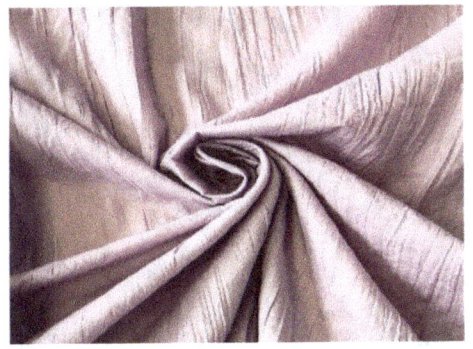

Suppose a coffee rotating in glass; it´s possible in CFD, analyze the surface and show a "heights map-field" or density, forming the spiral as a galaxy arms:

Few weeks ago, I went to Amsterdam; in Nemo (Science Museum) I have seen a tool for generating spirals from fluids in rotation:

Now. I want to create some simulations in order to know better, all that:

→ 1 Fluid in a disc in rotation; analyze the density variations, against viscosity, density, etc….

→ More than 2 fluids in in rotation; analyze the density variations, against viscosity, density, etc….

→ ….

One the goals, it´s may be, analyze the dependence of arms number against: Viscosity, density, velocity rotation.

These waves are zones low pressure and other´s high pressure. In the zones of high pressure, so with high density, it creates the stars, so the arms (with the help, as always, of the gravity):

If instability occurs in a rotating star cloud, a zone or zones can form that corresponds to a zone of higher density (stars tend to cluster together); as they rotate, the arms appear. The initial instability is responsible for the origin of the arms:

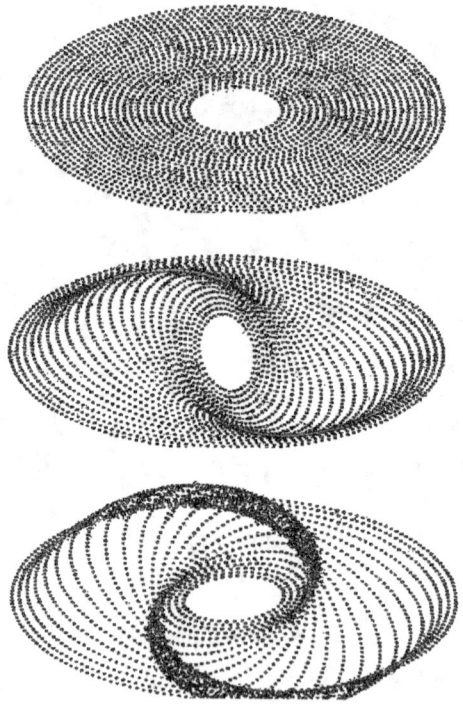

This origin of arms from gravity waves, is the similar to generation of traffic jam from a instability created for one car in acceleration, or brake (soliton).

→ Any explanation for the creation or formation of arms in a galaxy is based on instability.

In fact, if a cup of coffee is moved and the spoon touches the surface, this contact is an instability that develops in the arms as the coffee continues to rotate. This instability, as always, is the most important in arm generation.

For example, in a galaxy in bar, in the zone "A" (extremes) is more likely the generation of arm, caused by:

- Drag matter for the rotation.
- Generation of instability in "A".

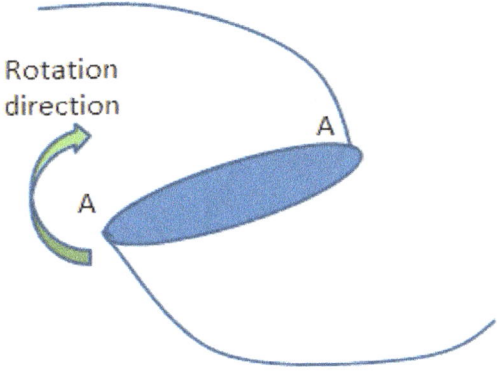

In other galaxy or better: stars accumulation, is possible the creation of arms from instabilities (6 in this image sample):

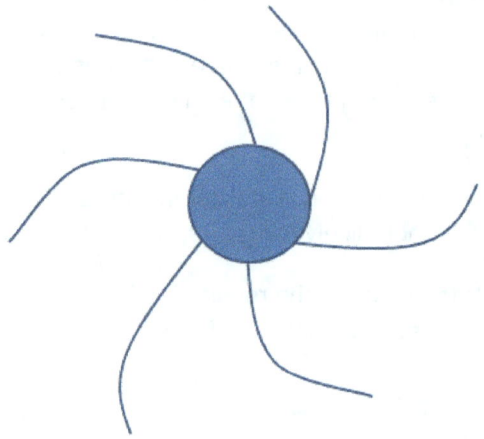

It´s not rare so, the result or conclusion (from Spyzer telescope and datas in Gaya mission of ESA) about the "break" in one arm of milky way: it´s perfectly possible a instability in this arm; this instability can produce this "break" or also the called "spur"; these spurs, are in 3D not only in the galaxy plane:

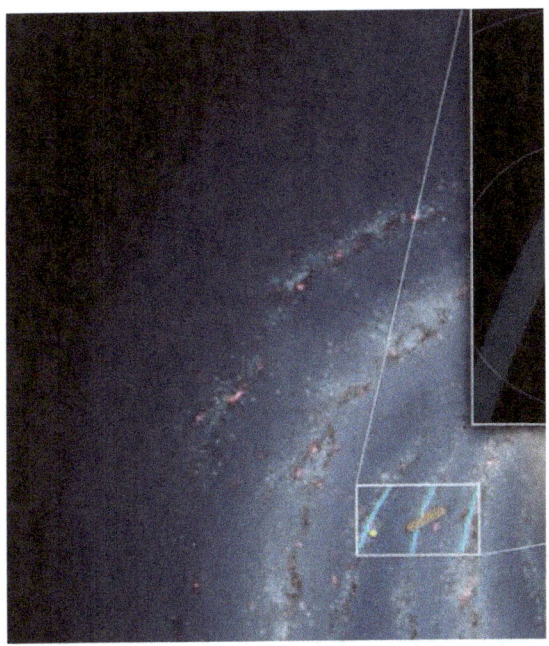

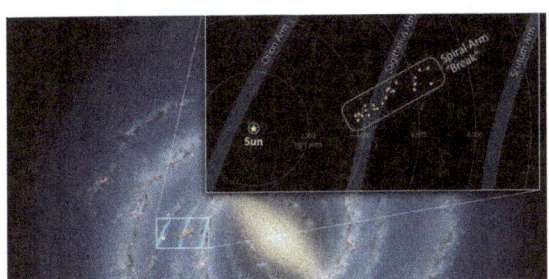

Other case: 2 "armless" galaxies can give rise to a single galaxy formed by several arms: 2 galaxies come together by the action of their gravities rotating around each other; by the action of the ram pressure, arms are formed in each of them which, when united, form the arms of the new galaxy:

Now. I want to create some simulations in order to know better, all that:

→ 1 Fluid in a disc in rotation; analyze the density variations, against viscosity, density, etc....: creating a instabilities.

→ More than 2 fluids in in rotation; analyze the density variations, against viscosity, density, etc........: creating a instabilities.

→ ....

One the goals, it's my be, analyze the dependence of arms number against: Viscosity, density, velocity rotation, instabilities (size, magnitude, number, etc....).

At certain heights in the atmosphere, there are different layers of air at different temperatures and therefore densities and viscosities. At that time, the interface, works like the free surface of the sea; waves with different frequencies and amplitudes are produced, producing different cloud geometries:

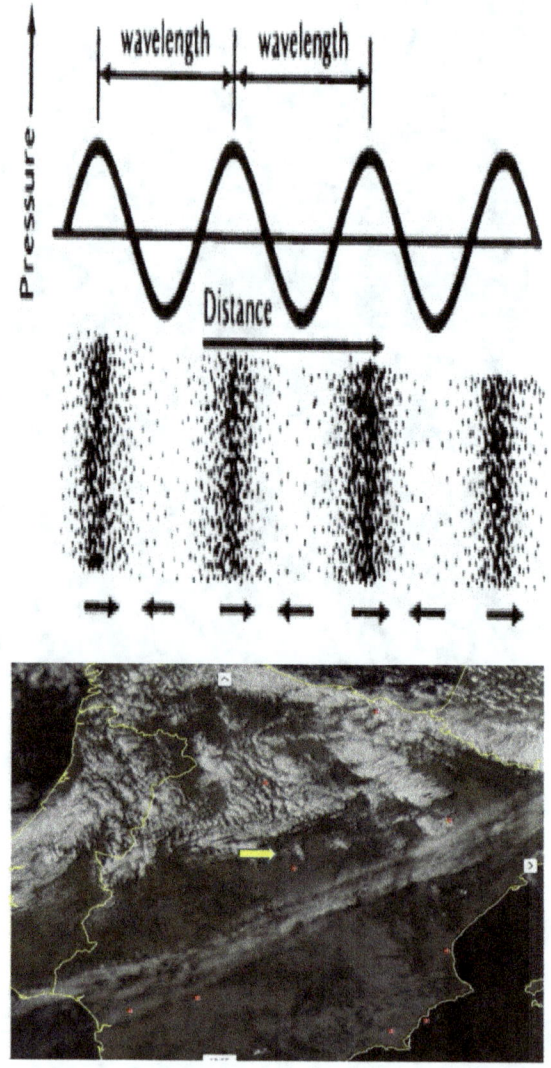

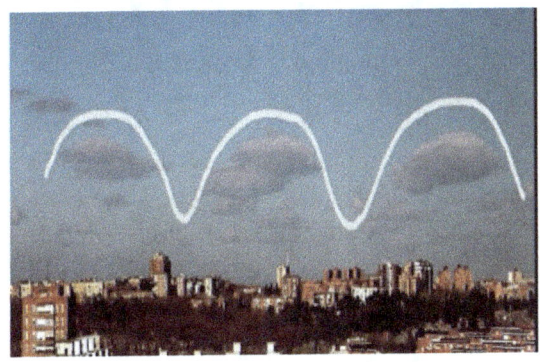

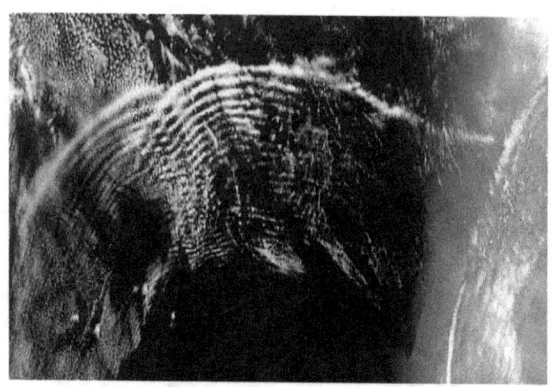

In the pick of every wave, there are more density, so in this zone, it´s possible to aggregate the matter.

These waves are density waves. Why a lot are shapes types of galaxies? Depend of initial matter distribution, including amount, velocity rotation black hole, dimensions of black hole or disturbance zone, environmental, etc.... The galaxies matter may be composed by baryonic matter but also may be, by Dark Matter transformed in baryonic matter.... (Mond theory ????).... It´s possible see the arms, through a density map or viscosity map-field generation (coffee or chocolate for example – is very complicate to see these densities variation or height variation - waves).

There are other methods for simulate a galaxy: creating a vortex as a galaxy. And more: the interaction between galaxies as a vortices.

It knows that a bullet before leaving for the rifle barrel rotates within the barrel. The goal is to provide rotation (inertia) to go "straight":

The bullet keeps rotating until it reaches its target.

The same happens when we throw a rugby ball: given its peculiar shape it must travel in this position with respect to the direction of the air.

Drag will be much smaller and therefore it may achieve greater distances. But to maintain this position it must rotate, and that's how it is thrown: in rotation:

Therefore, it can think of an alternative system, to generate high-energy rotational vortices, different to what we have seen up to now:

It can try to fold the airflow so that we keep the rotation applied previously.

This could be done by circulating air through a kind of auger, so that we force a spiral rotation:

When air circulates this device entirely we think that it will continue to maintain the rotation a certain distance after; however, this is completely wrong: just at the end of the auger its rotation is drastically and suddenly interrupted. It is always necessary to generate vortices due to a device that is based on pressure differences; why?

Because by this way, it produces a depression tube after system which suctions the air producing the vortex along the time: the translation of particle, produce the vortex.

We choice the method, placing a "wing" (there is a difference pressure so), in order to create a vortex:

First, it analyzes the next test:

- 1 Fluid.
- Wing in order to generate a vortex.
- Visualize the vortex, behind wing.

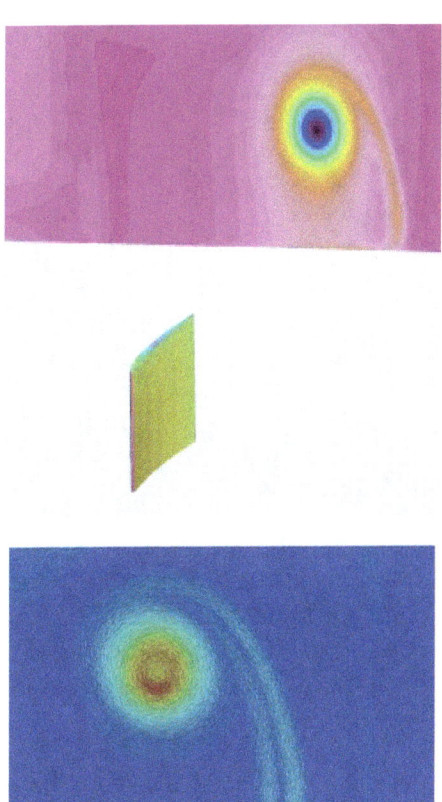

First image: blue-red → low-high pressure. Rest images: red-blue → high-low vorticity.

As a first test, good attempt....

It´s necessary so, more, in order to create and visualize better the arms.

There is other way for creating a galaxy:

Is an idea from Marco Fedi, Mario R.L. Artigiani (with some variations for improving it); is a body, in rotation, with low pressure in center:

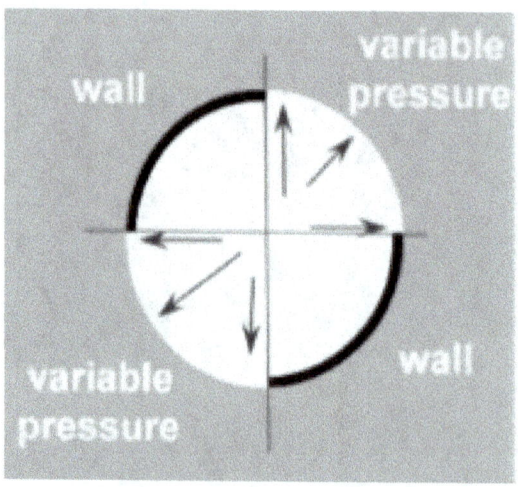

In the work of creators (Theoretical and CFD analysis of gravity and galaxy formation and rotation in a dilatant vacuum), the vortices is static; I propose simulate 2 or more vortices in order to analyze the interaction; for that, is necessary that the vortices, can move (in CFD that called 6-DOF or 6 freedom degrees).

It can make this body, with more than 2 walls, in order to create more than 2 arms for the galaxy. And more: in the center, it creates an outlet, and the mass trough that exit by outlet, is mass to add in center and so, more gravity or less pressure.

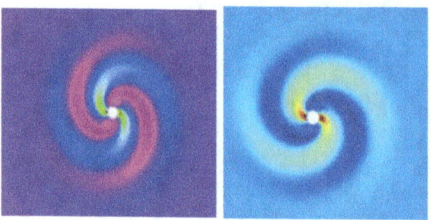

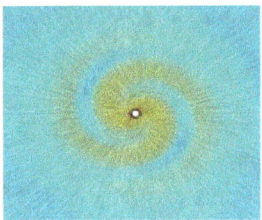

Also it´s possible, to generate 2 black holes, and analyze the rotation between them, generating the gravitation waves:

- Success horizon against size and mass.
- Rotation or not.
- Velocity and sense rotation.
- Etc….

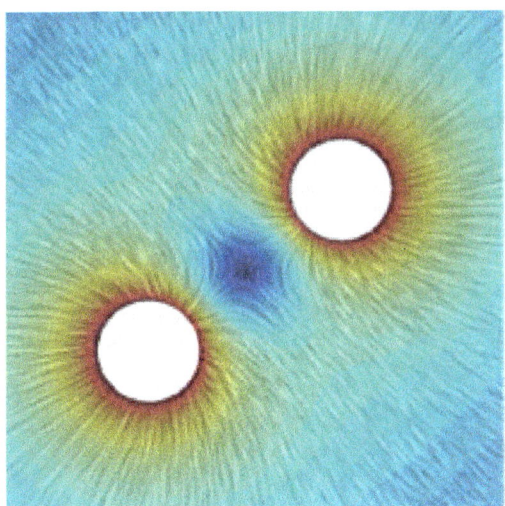

Phenomenons, properties, relation and interactions between, geometries, similarities:

All my life I have worked on aerodynamics applied especially to race cars.

For this reason, I have deeply studied the creation of vortices and the interaction between them.

The main objective is to apply this knowledge directly to the formation, evolution and interaction of galaxies. From this deep knowledge about vortices and turbulences in general, it´s possible to know phenomena that will occur in galaxies, because they occur in vortices: formation, evolution and interaction.

Next, there are several phenomena that occur in vortices and that although they have not yet been observed in galaxies, surely they will occur. It´s a matter of investigating such phenomena, and for that, I need time and means to do it.

a) LUMINOSITY / ROTATION VELOCITY

The galaxy matter in rotation, interaction with other matter, producing big greater densities zones. So, It´s possible to think, that if a galaxy have a greater rotation velocity, the interaction with the matter, also is bigger, so more stars zones formation. So finally,

If a galaxy have a greater rotation velocity, its luminosity will may be bigger. And that is true. Classification galaxies luminosity, against rotation velocity; Rubin in 1983:

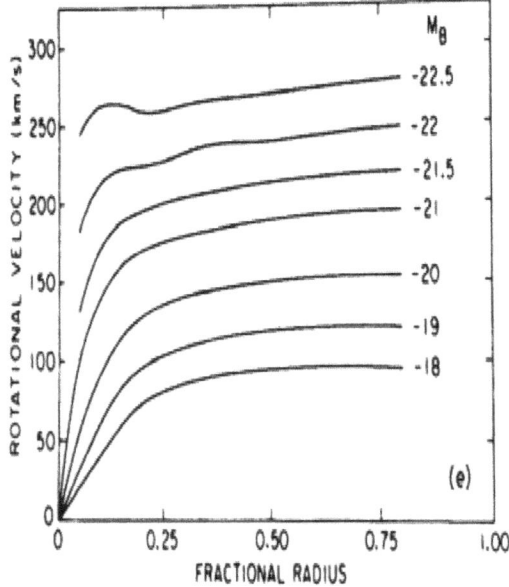

b) *LUMINOSITY / TYPE*

A galaxy has more luminosity, will must to be more arms. Classification galaxies Sa, Sb and Sc.:

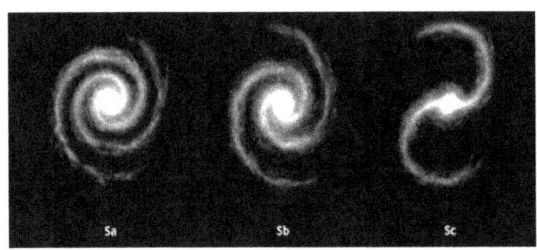

Rotation velocity "Sa", "Sb" and "Sc" galaxies.

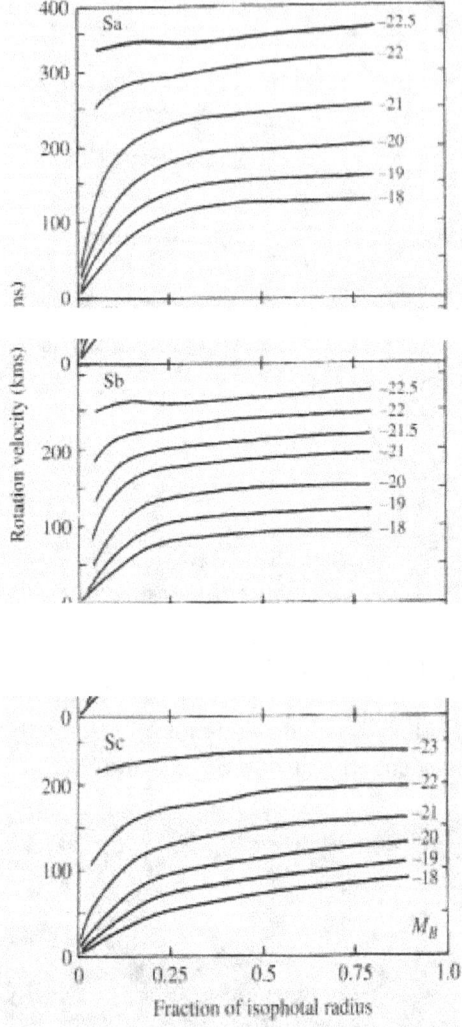

c) *GALAXIES TAIL*

When talking about the tail produced by a Galaxy on its way through the Universe, there are 2 types of Tails:

- The one produced by the stripper of the matter of the galaxy caused by Ram Pressure: high pressure.

- The galaxy, in its path, "cleans" the path, capturing matter and leaving a trace of low density: low pressure.

In one case, the Tail acts by repelling matter (if there is difference pressure and also may be attracting by gravity), and the second case, by attracting matter. The matter in a galaxy's path depends on whether the galaxy moves through more matter; in that case it will leave traces of itself combined with the surrounding matter due to Ram Pressure. In the case that there is hardly any matter in its path, it will drag what little matter there is or other types of interactions.

Sample about stripped galaxy and their CFD simulation:

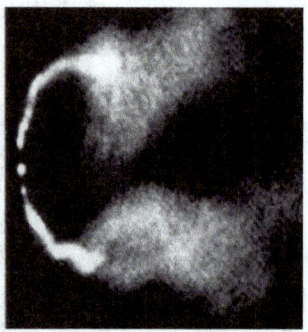

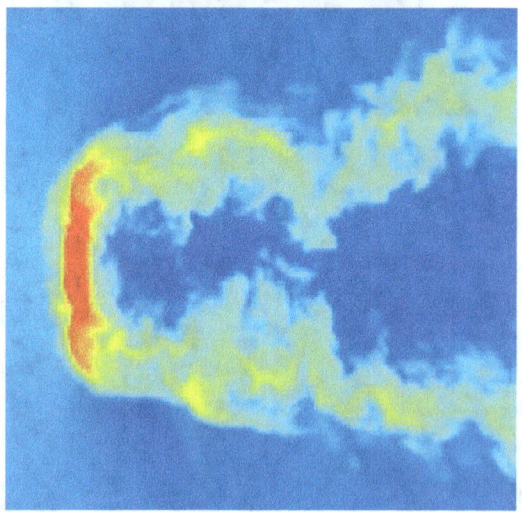

More about stripped galaxies:

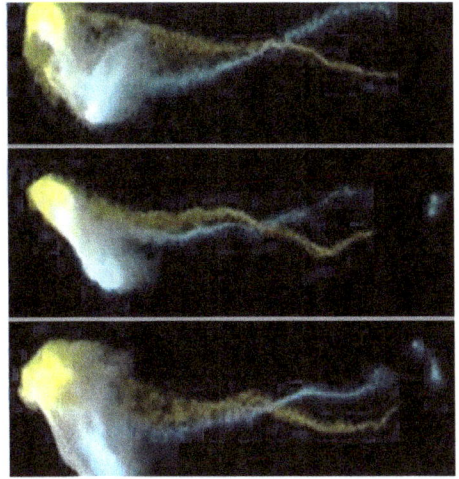

Simulation CFD by Rafael Ruggiero from Brasil:

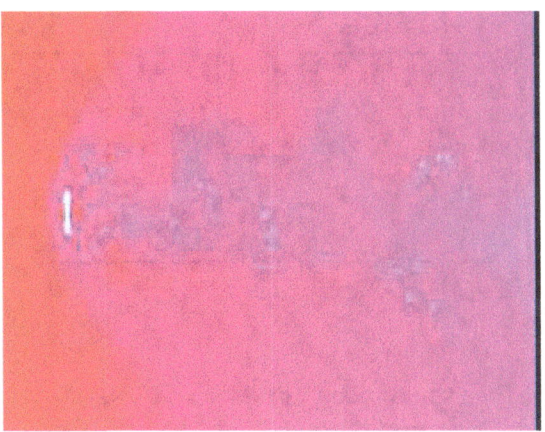

More CFD simulation by ([35] Elke Roediger, Marcus Brüggen and Matthias Hoeft):

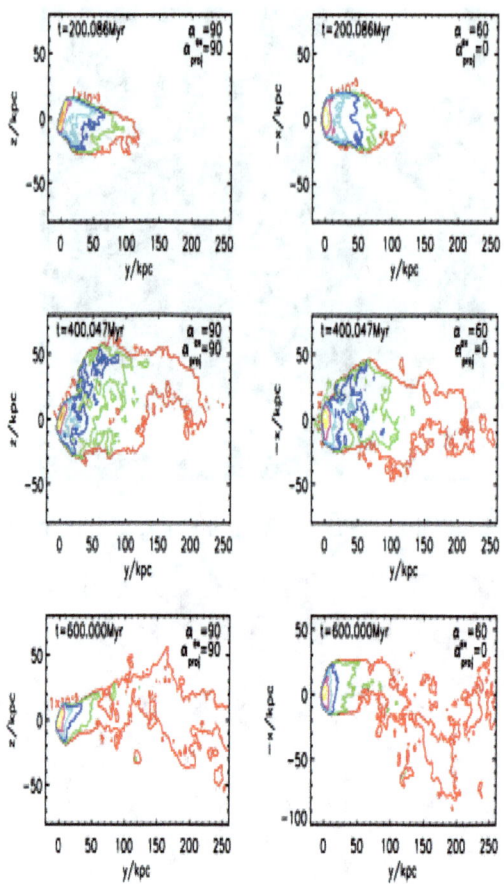

The same occur in Mallorca Island (tail sand formation) in storm:

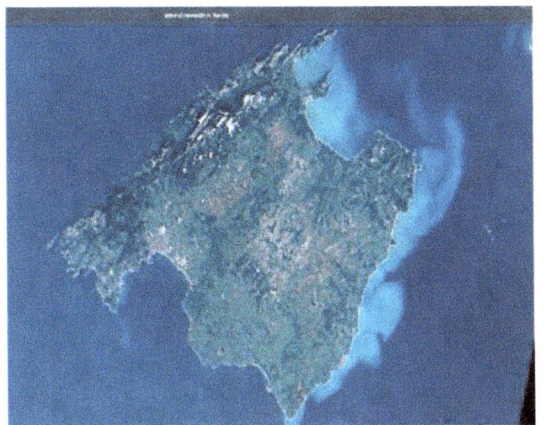

d) GALAXIES WITH LESS DARK MATTER

If there is little dark matter, the galaxy has fewer arms.

e) OLD GALAXIES AGAINST LUMINOSITY AND ROTATION VELOCITY

In the past (more density - far), a galaxy have more luminosity with less rotation velocity than today.

f) VELOCITY ROTATION AGAINST DISTANCE

Rotation velocity against distance:

Galaxy	RA hh mm ss	Dec dd mm ss	PA d	D Mpc	Type	m_{abs}	$\log D_{25}$	$\log v_m$
(1)	(2)	(3)	(4)	(5)	(6)	(7)	(8)	(9)
NGC 522	01 24 45.91	09 59 40.5	33.3	39.0	Sbc	−20.53	1.44	2.252
NGC 684	01 50 14.03	27 38 44.2	88.8	50.4	Sb	−21.53	1.53	2.368
MCG-01-05-047	01 52 49.01	−03 26 51.2	161.4	71.5	Sc	−21.77	1.47	2.410
NGC 781	02 00 09.02	12 39 21.5	13.0	49.8	Sab	−20.87	1.18	
NGC 2654	08 49 11.91	60 13 13.9	65.0	19.2	SBab	−20.09	1.62	2.295
UGC 4906	09 17 39.94	52 59 34.3	49.0	32.6	Sa	−20.26	1.30	2.228
NGC 2862	09 24 55.10	26 46 29.0	114.0	58.5	SBbc	−21.44	1.41	2.464
NGC 3279	10 34 42.61	11 11 50.7	152.0	19.9	Scd	−19.27	1.44	2.208
NGC 3501	11 02 47.35	17 59 22.6	28.0	16.2	Sc	−19.05	1.54	2.147
NGC 5981	15 37 53.55	59 23 30.9	139.5	36.1	Sc	−20.61	1.43	2.424
NGC 6835	19 54 33.09	−12 34 02.5	72.0	23.0	SBa	−19.55	1.38	1.803

D Mpc	Type	m_{abs}	$\log D_{25}$	$\log v_m$
(5)	(6)	(7)	(8)	(9)
39.0	Sbc	−20.53	1.44	2.252
50.4	Sb	−21.53	1.53	2.368
71.5	Sc	−21.77	1.47	2.410
49.8	Sab	−20.87	1.18	
19.2	SBab	−20.09	1.62	2.295
32.6	Sa	−20.26	1.30	2.228
58.5	SBbc	−21.44	1.41	2.464
19.9	Scd	−19.27	1.44	2.208
16.2	Sc	−19.05	1.54	2.147
36.1	Sc	−20.61	1.43	2.424
23.0	SBa	−19.55	1.38	1.803

Analyze now, "D" and "V" relation. Relation between Rotation velocity against distance:

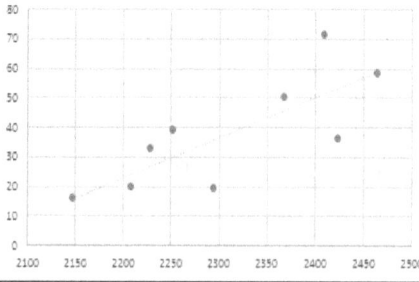

There is a relation lineal. Ok…………..

g) *VELOCITY ROTATION AGAINST MASS*

Velocity rotation, against galaxy mass:

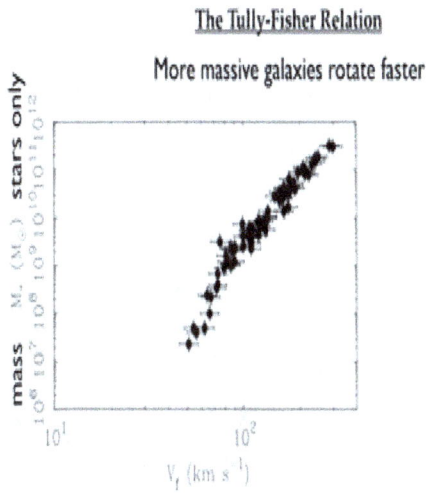

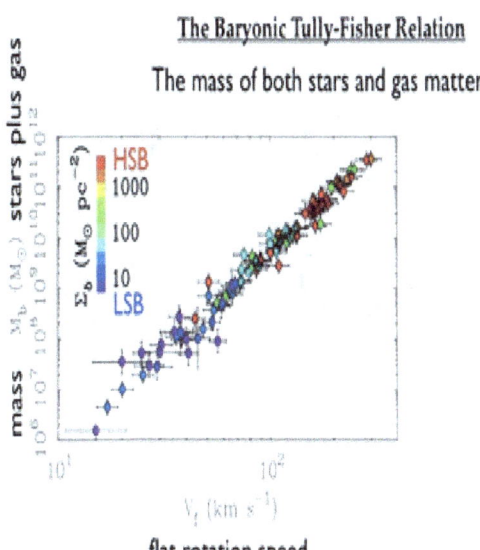

h) AGE AGAINST ARMS

There are a relation between galaxy age and arms (number arms, luminosity and rotation speed). That is: if the galaxy is older, the galaxy has more arms (also in general, obviously): Triangulum 2.38 to 3.07 Mly, Pinwheel 20.9 ± 1.8 Mly, M51 37 Mly:

That is "normal" because if it´s older, have more time to "work" with the environmental. A young galaxy, has less arms, is irregular and diffuse:

i) INTERACTION BETWEEN

A) INTERACTION AS A VORTICES WITHOUT GRAVITY

In the sky, it´s possible to see, a lot samples of collisions:

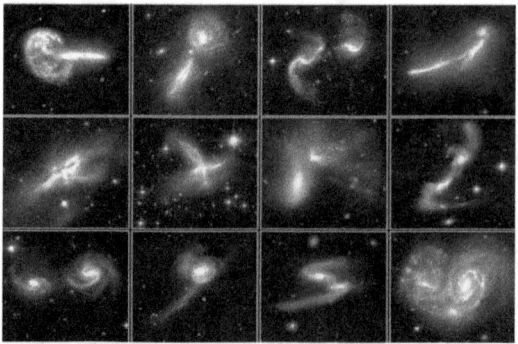

When two galaxies collide, collide also the dark matter.

It´s possible to see the same between hurricanes (analogy with fluid theory or geometries patterns):

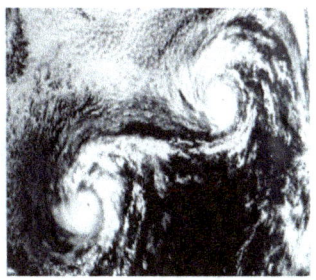

So, it can think about (evolution and combination), as an interaction between fluids vortex. The interaction between vortices is some think very important and complicate:

These unions or alteration, depending of intensity (vorticity) of vortex; if there is one vortex, bigger than other vortex (red and green), can produce that: (each horizontal file correspond one context or size):

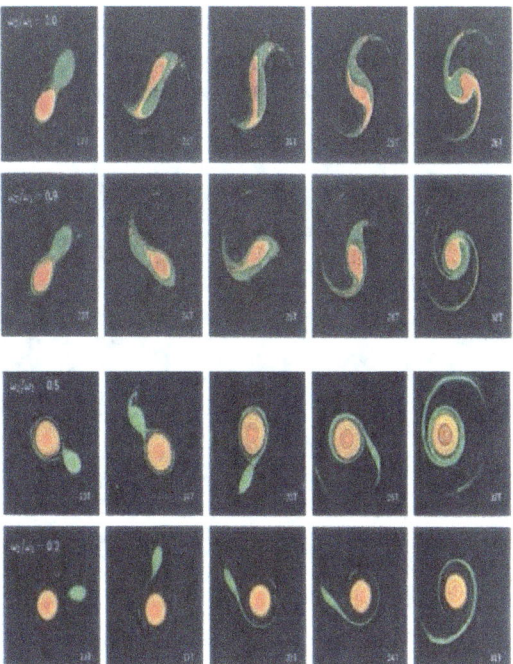

Also depend of densities, viscosities, temperatures, rotation direction, velocity rotation, size, etc.... and other´s paths as a way of other galaxies (depression tubes):

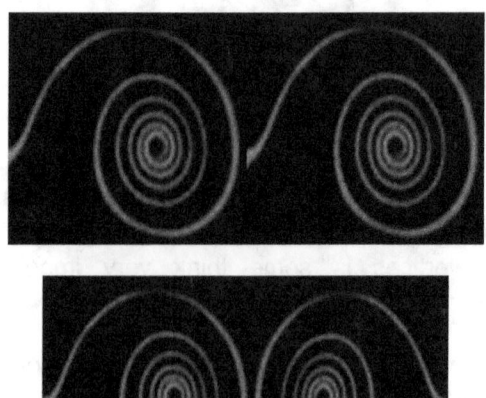

The study of these interactions between vortices, is typical in aerodynamic work about Race Cars; is important create vortices, but is more important, their interactions. Vortices in front wing race car Formula 1:

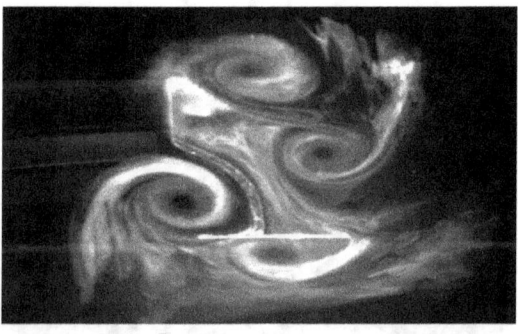

In this typical case, there are lot samples of interaction between vortices with the same or different turn sense, sizes, velocity, etc.... (the same in galaxies). It could simplify the problem of the interaction of 2 or more galaxies, assuming it work in 2 dimensions, but this is not real. Galaxies are 3-dimensional and so is their location. Interactions are not executed in a plane. Galaxies, interact with each other, giving rise to many different geometric shapes, depending on several factors:

- speed of rotation.
- size.
- quantity of mass.
- 3D location.
- direction of rotation.
- direction translation.
- Etc.....

On the other hand, it is possible perfectly, to simulate the interactions between galaxies and also the formation of galaxies, by CFD Simulation:

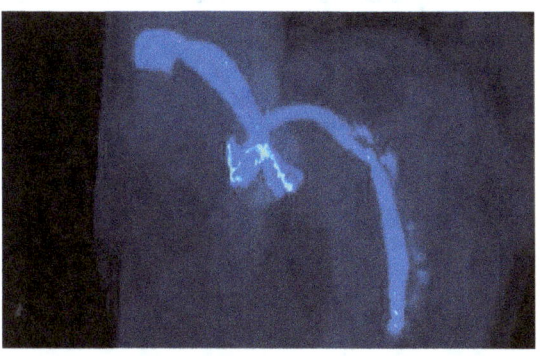

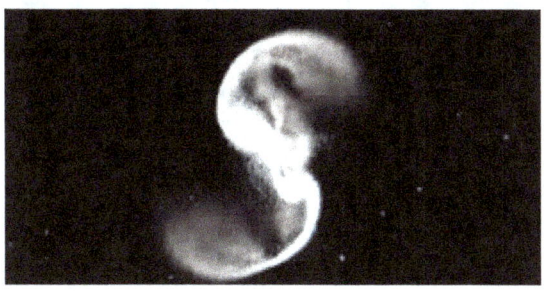

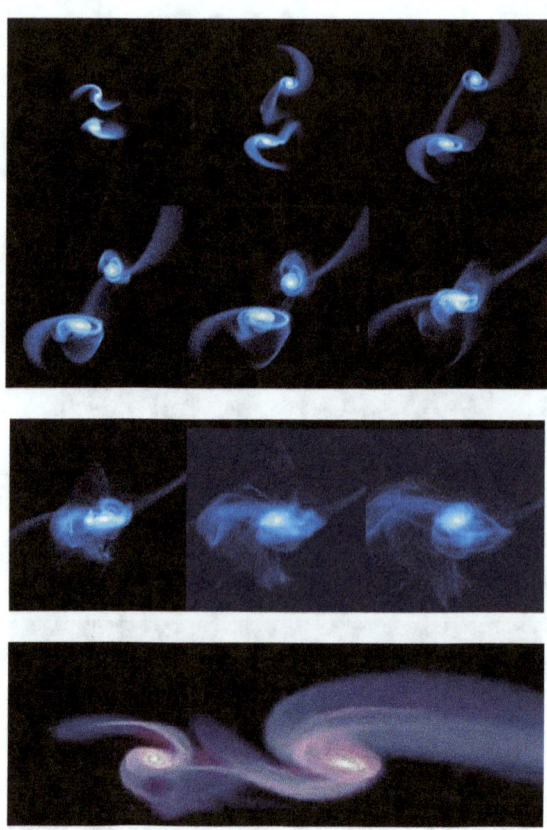

**B) SENSE ROTATION – 1 ,
INTERACTION**

Article ([37] J. H. Lee et al):

The sense of rotation of a galaxy is influenced by the displacement of its companions, even the farthest. This is revealed by the CALIFA galactic survey data used by a group of astronomers to carry out the study.

In principle, distant companions, located millions of light years away, should show little influence on the shape and rotation of the central galaxy, but a recent study indicates that the direction of rotation of a given galaxy depends, in effect, on the average displacement of its neighbors, including those located at long distances.""

It can see another very illustrative effect of the importance of the density and viscosity and this last Article.

Let's think of a submerged pendulum. It makes it swing.

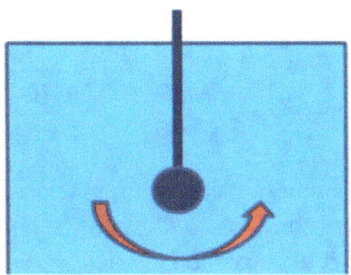

It will be able to see that the pendulum will stop oscillating almost immediately. This is due to the opposition of the water molecules which act on it. In fact, the more density/viscosity the fluid has (less compressibility), the less time the initial oscillation will take to stop.

Now, let's think of two identical pendulums immersed in a fluid and with opposed oscillations.

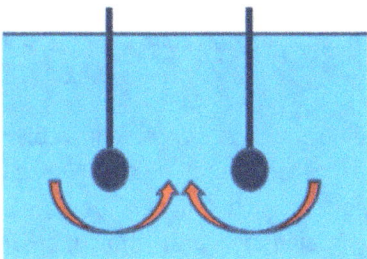

After a short time, both pendulums will oscillate in the same direction and with the same frequency!!!

Why does this fact happen?

Because the density/viscosity of the fluid, because its variations and the forces transmission trough the particles. On the moon, this wouldn't happen, due to the air absence.

This morning, walking with my mother, she said to me: we're going with the wrong step. In other words, we don't have the same footprint in the same instant. This makes the walk "unstable" and unpleasant....

C) SENSE ROTATION – 2, AS A GEARS

For 2 nearby galaxies (can also occur in more distant galaxies), they rotate in opposite directions. This is because, in their rotation, each galaxy drags its companion, like 2 gears (Magnus effect):

D) RESONANT ORBITS IN GALAXIES

Article: Francesca Fragkoudi: orbits of stars in a Milky Way-like barred spiral galaxy; these are resonant orbits extracted directly from one of the Auriga cosmological simulations, and shown in an inertial and rotating frame of reference:

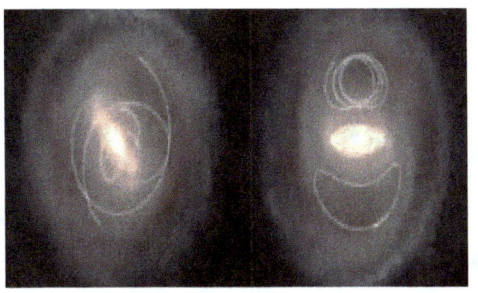

E) MAGNUS EFFECT

There is an effect associated with the rotation of a galaxy; it is the Magnus effect. A galaxy is a group of matter or energy in general, rotating more or less compactly; this causes the galaxy to have a friction with the matter that "may exist" around it, in its path: more matter or density, more friction and more Magnus force; this density is energy density.

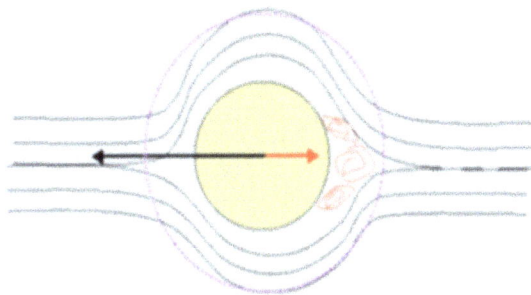

It can observe that there is no difference between the lines at the top and bottom part. Meaning: there is no difference of pressure. Wouldn't this also depend on the Reynolds number? For some transition Reynolds, we would have temporary behaviors, non-symmetric solutions which are known as the von Karman Street, but this is not the case that concerns us to understand the phenomenon.

Now, let's suppose that the same ball is rotating. There is, therefore, a perceptible difference of pressure on the top and bottom of the ball. The air goes from the left to the right in the image:

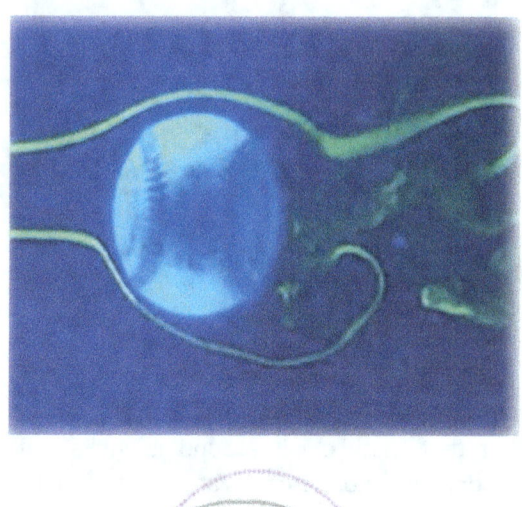

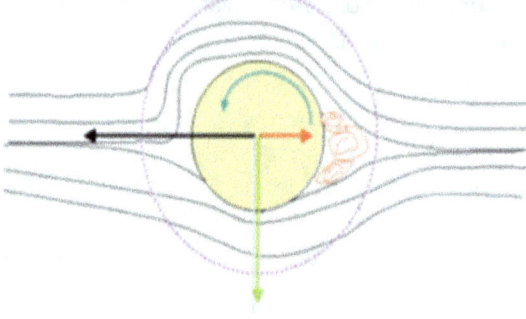

If it turns the ball in the direction indicated with the green arrow and if there is this difference of pressure, the ball will tend to go down: there is lower pressure below because the flow circulates quicker.

Depending on the direction in which the ball rotates, it will go up or down. In other words, the resultant force will have a determined direction, depending on the direction of the ball.

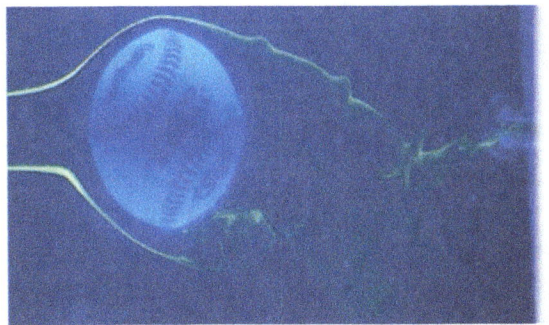

This friction between galaxy and matter, together with the direction of rotation, produces a force that makes the path tend to follow a specific path:

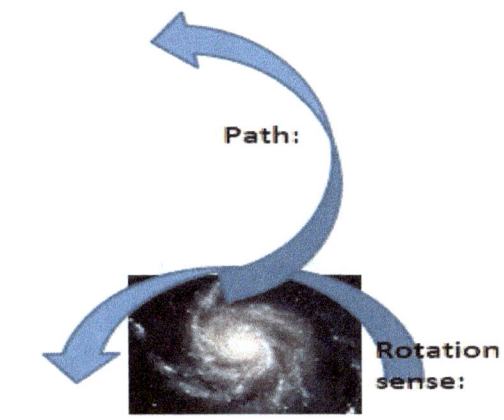

For example, in moon, is not possible doing an effect with football:

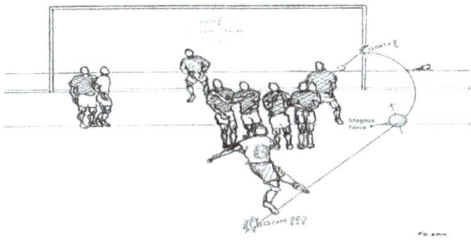

This effect is applied to low pressure tubes as a galaxies paths. So is very important in order to know the full interaction between galaxies.

F) GALAXIES: FULL INTERACTION

It´s a set and consequence of all effects before. In this Article, it analyzes Navier Stokes Equations, from the simplest versions, analyzing term by term and ending with a full version introducing even the electromagnetic effects.

To "really" analyze the interaction of galaxies or the creation or formation of the Spyder web of the universe, it´s necessary to apply the more complicated Navier Stokes equations, since they reflect perfectly, what happens in reality.

Each galaxy or group of galaxies, like tail, leaves a path altered in density or pressure.

These "virtual" pathways are there and exist, but they are very difficult to locate and know but they are there, and they determine and modify the path of other galaxies that approach these pathways.

This path "virtual", is a factor very important in galaxies dynamic: the dynamic galaxy, produce an alteration in environmental.

Galaxy tails (high or low density), also evolve, move and change. The reason is that the environment affects them: other tails, zones of low and high density, etc....

The evolution of galaxies is based in:

- Viscosity.
- Interaction between Tails.
- Gravity.
- Others forces….

Here, an image very important: Star bridge in Magellanic clouds:

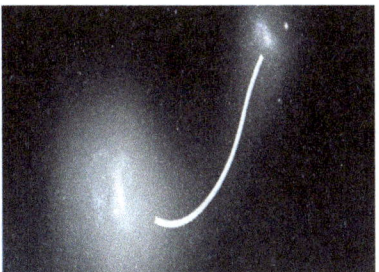

→ It´s possible to know what is the path or direction of displacement of each galaxy, studying the star bridges, tails, etc....

About this last affirmation or possible research, there is an Article very recent ([37] Ekta Patel and other´s):

"With the release of Gaia DR2, it is now possible to measure the proper motions (PMs) of the lowest mass, ultra-faint satellites in the Milky Way's (MW) halo for the first time. Many of these faint satellites are posited to have been accreted as satellites of the Magellanic Clouds (MCs)"

From this Article, it´s has a table of position and velocities of galaxies set, in order to analyze the interactions (in the past and the future) or paths between:

	X	Y	Z	V_x	V_y	V_z
	[kpc]	[kpc]	[kpc]	[km s^{-1}]	[km s^{-1}]	[km s^{-1}]
Aqu2	28.71±1.23	53.16±1.77	-85.98±2.87	91.31±239.21	250.76±212.3	130.49±166.0
CanVen2	-16.37±0.22	18.58±0.51	158.67±4.32	-0.66±162.9	-203.05±150.42	-70.09±16.93
Car2	-8.3±0.0	-34.54±0.8	-10.65±0.25	134.12±11.0	-287.58±4.14	134.95±13.02
Car3	-8.29±0.0	-26.6±1.24	-8.06±0.37	-10.7±18.9	-151.85±8.41	356.05±25.9
Cra2	10.3±0.88	-81.23±3.86	75.13±3.57	-34.4±35.2	115.88±21.41	2.83±19.96
Dra2	-10.57±0.04	15.58±0.28	14.61±0.26	22.54±22.16	100.31±22.35	-341.04±25.48
Hor1	-7.16±0.1	-48.01±4.36	-67.91±6.16	-20.24±30.24	-150.18±45.34	152.34±32.09
Hyi1	1.87±0.19	-19.59±0.36	-16.48±0.3	-144.15±6.58	-178.7±8.73	288.26±8.57
Hya2	47.82±3.06	-117.14±6.39	76.34±4.17	-165.16±302.26	-92.01±257.22	208.27±275.52
Phx2	25.47±3.14	-24.81±2.31	-71.85±6.69	-67.68±48.82	-165.47±54.59	162.72±31.4
Ret2	-9.63±0.06	-20.38±0.96	-24.14±1.14	19.92±12.38	-96.74±17.42	218.24±14.63
Seg1	-19.38±0.98	-9.47±0.84	17.67±1.57	-98.19±18.34	-205.06±38.14	-35.49±22.9
Tuc3	0.79±0.41	-8.95±0.4	-19.03±0.85	23.48±5.94	146.27±8.05	185.68±5.69
Car1	-24.72±0.6	-94.62±3.48	-39.26±1.44	-36.84±18.42	-50.55±8.51	149.23±20.28
Dra1	-4.15±0.32	64.88±5.0	45.01±3.47	54.29±13.85	4.15±8.25	-151.78±11.73
Fnx1	-39.58±0.57	-48.15±0.87	-126.93±2.3	38.14±22.76	-107.56±21.25	76.0±9.72
Scu1	-5.22±0.19	-9.59±0.6	-84.12±5.26	16.93±12.7	175.89±16.04	-96.14±1.87
UMin1	-22.16±0.71	52.0±2.68	53.46±2.75	-4.26±10.75	46.77±10.69	-148.2±10.61
LMC	-1.06±0.33	-41.05±1.89	-27.83±1.28	-57.60 ±7.99	-225.96±12.60	221.16±16.68
SMC	15.05±1.07	-38.10±1.75	-44.18±2.03	17.66±3.84	-178.60±15.89	174.36 ±12.47

There is a "particular" type of vortex interaction, called Crow's Instability. I don't know why the "fashion" of naming many things, but Crow is basically, or primarily, a case of 2 counter-rotating "equal" vortices coming from the vortices of the wing tips of an airplane.

Extending this "definition" further, what is attempted in this Article is to know or model the interaction of various vortices, to know their past and future paths, knowing initial data (as in the case of Galaxy paths).

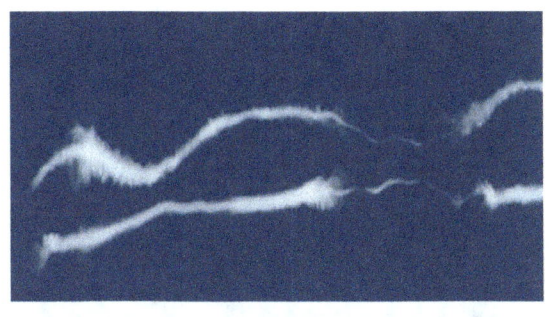

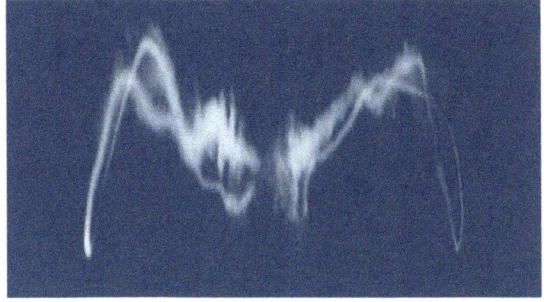

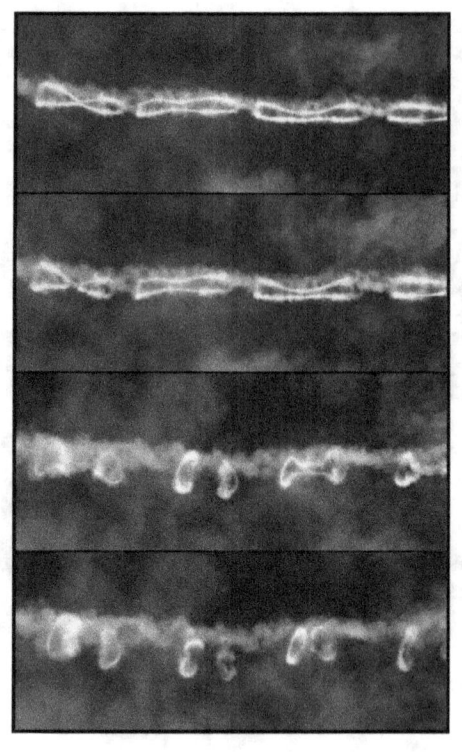

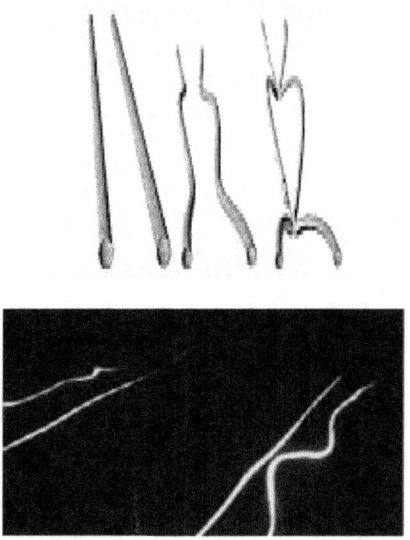

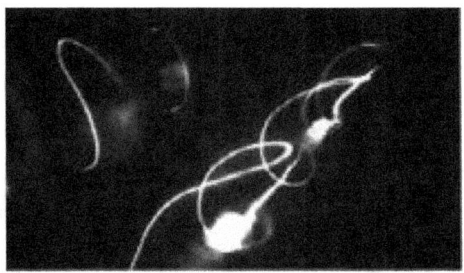

As it sees before, in Race Cars Engineering, we engineers have been studying for many years the interaction between Vortices created by ailerons (images of rear/front wing, by Jonas Pangerl and Marius Imiolczyk).

Meteorologists have assigned names (Crow instability) to these well-known vortices. They are still low-pressure "tubes" in which particles rotate.

The CFD techniques have a very useful tool, called "Core vortex", which perfectly points to the heart of the vortex, so as to analyze its path.

Vortex paths depend on many factors:

- Speed of rotation.
- Size.
- Sense of rotation.
- Density, Viscosity, humidity, etc.
- Of the paths of other vortices.
- But above all, pressure variations.

In this article, the CFD is pointed out as the tool capable of generating galaxies and observing their evolution and interaction with other galaxies, analyzing the interactions of the vortex cores of the vortices.

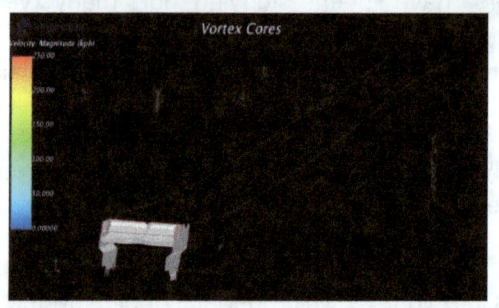

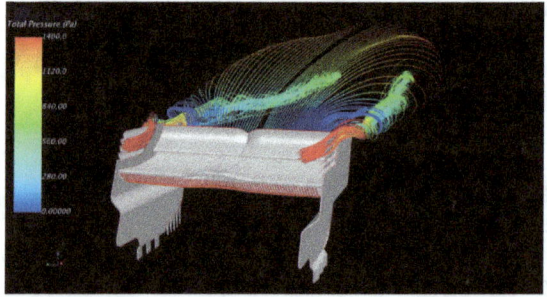

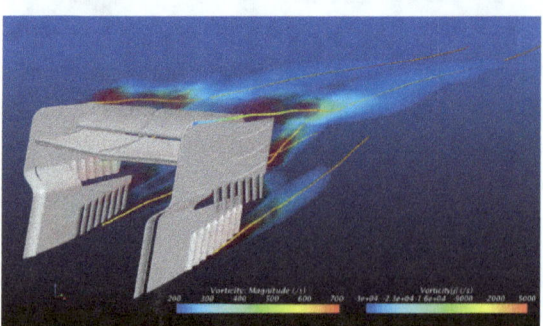

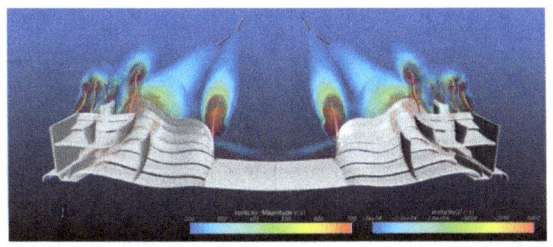

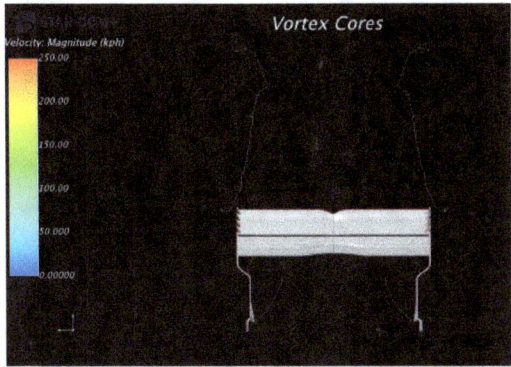

About the interaction between galaxies, it´s very interesting to see to movement of tails; that it´s normal, even necessary; in fact, the radio emission in center Abell 1775 (galaxy), observed by LOFAR (red) and placed in a optic image (Lofar / Pan-Starrs / Botteon et al. 2021):

In ([42] Peng Wang, Dr. Noam Libeskind, Sarah Hönig): "By mapping the motion of galaxies in huge filaments that connect the cosmic web, astronomers at the Leibniz Institute for Astrophysics Potsdam (AIP), in collaboration with scientists in China and Estonia, have found that these long tendrils of galaxies spin on the scale of hundreds of millions of light years. A rotation on such enormous scales has never been seen before. The results published in Nature Astronomy signify that angular momentum can be generated on unprecedented scales".

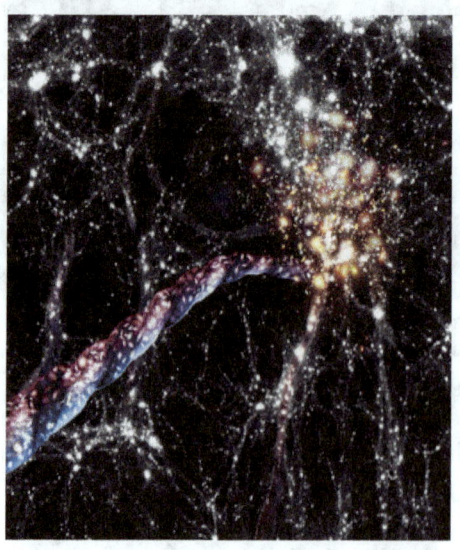

When a galaxy advances, it leaves, as seen in this work, a low-pressure trail. The low-pressure or low-density path will be larger (less density) if the galaxy:

- Is large.
- Has a lot of velocity.

When another galaxy encounters this low density tube, it starts to spin around the tube; this causes the paths of both galaxies to merge and grow; this can happen to many galaxies, so the pressure tube becomes very large. Many galaxies rotate around and inside the low-density tube; this effect is discussed in this article.

About these last points (A to F and more, I want to make some simulations CFD in order to corroborate every asseveration in every point:

- Create 1 or more vortices for analyzing:

"A" to "F".

- **Spyder web or geometry universe in high scale**

The Background Microwave Cosmic (BMC), is a map of temperature variation in a distribution of mass in early universe (colors scale blue-red / low-high density or temperature):

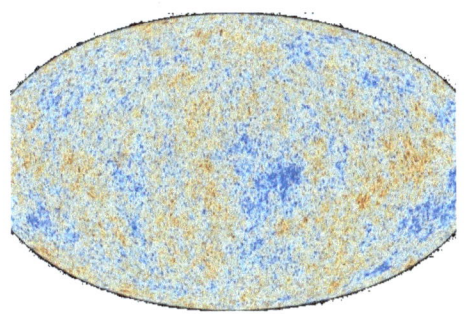

Why this special final distribution or dynamic? This distribution of zones with more and less density is normal in the Nature. All explosions or blasts for example, not have an equal matter in any direction or point, fragments distribution, density or temperature (sun surface, supernovas, atomic bomb, nebulae, etc....):

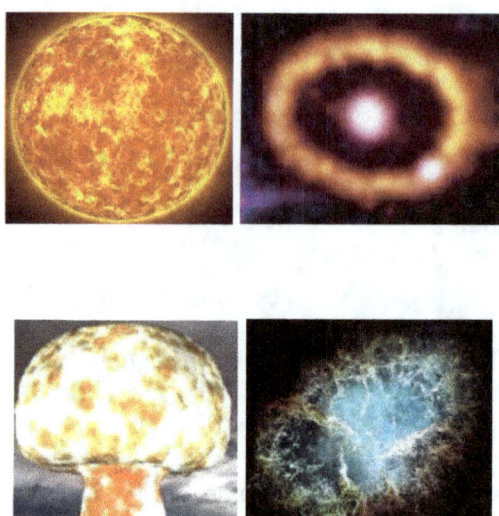

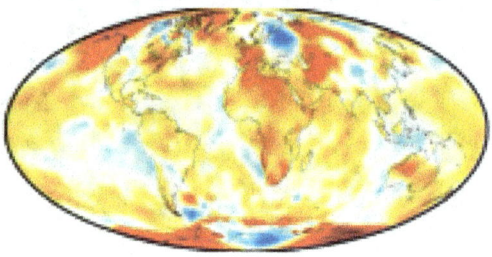

The same occur in temperature in earth surface: there are little's variations:

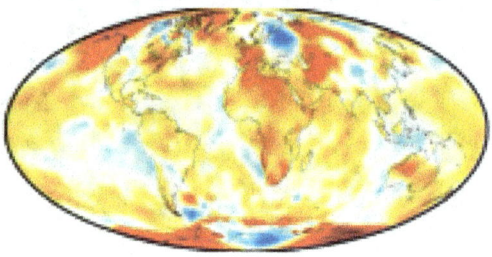

2 ways or numeric models, for explaining this special structure:

→ First way (simple method): it supposes that the Universe work as a fluid with density-viscosity (galaxies, environmental, etc):

These densities-viscosities variations (CMB), origin in the future, the different galaxies cluster and matter distribution in large scale. In fact, from this BMC as a boundary condition, is possible simulate the evolution of universe: the result is very similar to universe observable today:

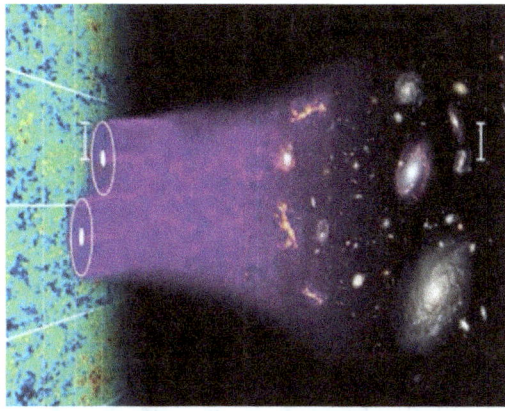

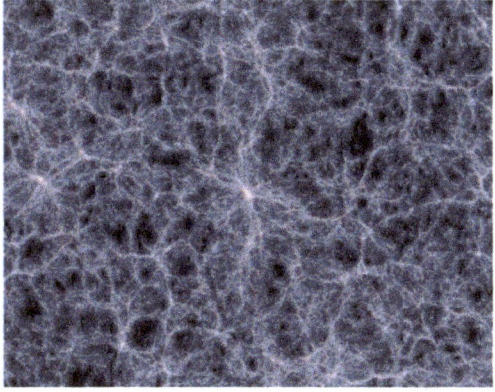

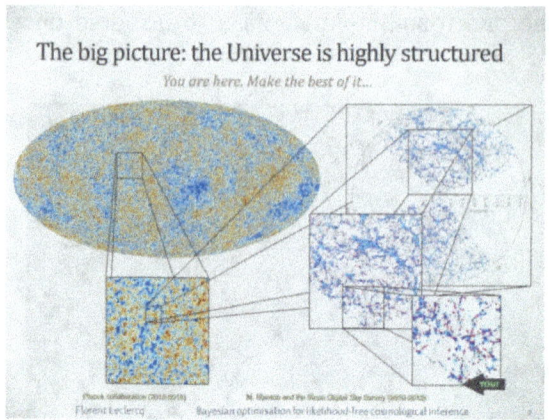

The big picture: the Universe is highly structured

You are here. Make the best of it...

→ There are other way for creating the universe in high scale (more complex but more accurate):

For the formation of the web spyder or large scale structure of the universe, in a numerical modeling can be applied, Navier Stokes: with these equations, incorporating viscosity and external forces (such as gravity, magnetism, etc), it is possible to simulate the large-scale structure of the universe.

The speed of light, like the speed of gravitational attraction in the universe, depends on the environment. The speed varies depending on the density, viscosity, magnetism, etc.... This difference of "action" between particles, called now, Viscosity or difference of reaction time, is what produces the aggregation of matter-particles, forming the cosmic web.

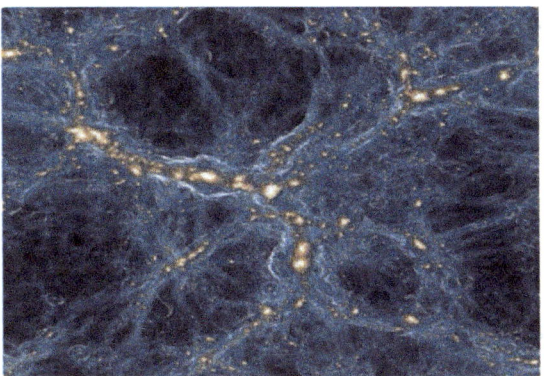

This structure called "Spyder web", is a Multifractal ([16] Vicent Martínez García).

It can also find this special geometric structure, in other field:

Abstract:

Predicting what is going to happen is very tempting, and it is something that everyone would like to be able to do, in an easy and above all reliable way.

Knowing a priori if it is going to rain in the next few minutes, while it is raining cats and dogs while having this desire is not something too difficult. However, knowing if it will rain in 46 and half days is not difficult at all. To know, for example, if tomorrow will be the day before, is easy, but it is not easy to know where a stone thrown with all the strength will fall.

In the determination or knowledge a priori of something, there are basically two aspects involved:

- The nature of the phenomenon itself.
- The time series we are using.

The essence of chaos is the scarcity and simplicity of laws that govern evolution; thus, it is easier to determine the evolution of the stock market than how a football will move; on the other hand, the more data we have about the evolution we want to simulate, the easier it will be for us and above all, the more reliable it will be.

In every chaotic process, there is a certain geometrical structure that is either visible or not observable: Fractal structure or geometry is the science that studies the representation of chaotic dynamics; we find fractal geometries in nature itself:

- Contours or borders of countries and coasts.
- The structure of a tree.
- The network of a person's blood circulation.

But also in representations of certain phenomena:

- The classic problem or simulation of a pendulum and 3 magnets of different colors.
- Etc....

One of the objectives that could be achieved by determining whether or not certain geometry is multifractal is to differentiate or classify phenomena according to their distribution of fractal dimensions.

Chaos:

The essence of the so-called "chaos" and of fractal and multifractal geometry, is precisely the scarcity and simplicity of laws: simple and scarce laws, originate structures of all kinds extremely complex and unpredictable; we only have to think that only one force, perfectly quantifiable, is responsible for the structure of the universe, with its galaxies and groups of galaxies, planets and planetary systems, etc. The fewer and simpler the primary forces that manage or describe a phenomenon, the more disorder and chaos there will be with respect to time and also space: for example, it is easier to predict the evolution of the stock market, than the evolution or dynamics of a fluid, the variations of a pendulum or simply where a stone thrown by us might fall. But also: the easier a problem seems to us, the more difficult it will be to find its foundation, laws and interdependence between factors, and the more difficult it will be to simulate its evolution in time.

Without chaos, there is no evolution, life or change of any kind.

This leads us to the statement of a new vision of chaos: the more unpredictable and the more difficult it is to solve a problem or a phenomenon, the simpler and more straightforward will be its foundations and the laws that govern it.

There is a very widespread concept or idea, which is directly or indirectly related to the concept of chaos or disorder; such idea is the so-called "butterfly effect". This idea is basically false or at least erroneous in its traditional concept; 2 events, causes or effects, are not necessarily united nor is it necessary for them to be dependent, even though it is said that they are in a very distant time or that their effects or inter-effects are appreciable in an infinitely small measure; it is essentially and conceptually false and erroneous. It becomes necessary, therefore, to define related and unrelated events.

How does the science of chaos help to explain such complex and different processes?

In recent times, the idea has been emerging that introduces the notion of chaos as a central element in research and in almost any scientific explanation or response. The idea of chaos and apparent disorder is a tool, which helps to understand certain phenomena that so far are almost inscrutable.

"The cosmos par excellence is the world, the absolute whole that contains all the partial alls," says Marcel Conche in "The Notion of Order," and disorder, adds Georges Balandier, "cannot appear except as a rupture of unity, of general harmony, and as an obscuring of purpose.

Chaos and disorder, as challenges to scientific thought, invite us to find the regularities of the irregular, the determinations of the indeterminate, the order of disorder.

Incongruent perhaps? We do not believe so. In recent times, this science that studies the relationship between chaos and the perceptible and non-perceptible world has been identified as "chaology". For Balandier, in his book: "The disorder", the chaology "seems to be concerned, at first, only in the curiosities or the deviations of the illusion in benefit of a science that has become strange. For her, triviality becomes mystery. The leaking tap is no longer a small domestic matter and a source of irritation, but the occasion for an erudite observation, made over the years, that makes this anomaly a kind of paradigm of chaos. The water of a waterfall, with its fall in layers, its dispersion in a multitude of droplets and its subsequent circulation towards the errant current, manifests a higher level of this complexity with a disorderly rhythm. Cigarette smoke, the companion to the wanderings of the spirit, which first rises in a straight line and suddenly twists and composes moving figures, suggests the presence of a similar phenomenon. Above, very high up, the marvellous clouds run, they construct celestial landscapes, mobile and always changing, chaos with which dreams are linked; but the new discipline wants to force its mystery, to find the answer that will make the forecast of the time beyond the immediate less fallible.

Chance is a determining factor in the manifestation of diverse phenomena and processes of the universe, and yet these are not as random as they appear or simulate:

Mitchell J. Feigembaum, states that "we are full of chaos", beauty is "essentially chaotic", the shape of the clouds is also chaotic. The science of chaos is for him "the study of disorder, of the irregular behavior of deterministic things, those that you know how they behave from one instant to another, and yet their displacements become irregular, erratic, and give the sensation that they are produced at random. And in reality, what happens is that they don't happen by chance.

Julieta Fierro, comments that "in particular the transitions of the particles can be studied more adequately by resorting to chaos theory.

Thus, it is possible to conceive of other universes, parallel to ours and totally incommunicado, each with its very peculiar physics determined by chaos.... Thus the universe possesses order and disorder, cosmoses that form and expand and can give origin to symmetrical bodies like the galaxies with stars, ringed planets and life. The presence of chaos at all scales of the universe implies a great diversity of possibilities and, therefore, one or several universes with enormous potential to create diversity". In biology and medicine, chaos offers answers to problems such as issues related to blood circulation: turbulence breaks up circulatory and cardiac regularity. In the treatment of epilepsy, the creation of electrical turbulence in certain areas of the brain may be able to block or mitigate the attacks or convulsions caused by this disease or disorder. The applications that chaos can offer are immense: economy, stock market, political transitions, evolution of social conflicts, human relations, negotiations, etc...

Self-similarity:

One of the best-known geometries, perhaps because of its beauty and transcendence, is turbulence itself. Leonardo da Vinci (1500 A.D.), already described, in a truly masterful way, as it could not be less, which is the structure of a turbulence: "the small eddies are almost numberless, and large things are rotated only by large eddies and not by small ones, and small things are turned by both small eddies and large".

Who has not experienced, tested and observed the existence of personal problems, among the members of a group or team, and likewise, other types of problems in a subgroup of the main group, and so on? We have all dreamed at some time, that our society and ourselves, we are not more than integral part of another humanity or society formed by giants, as they are for example, the bacteria for us.

Something similar thought the ancients, when they said that the Milky Way in the night sky seen from the earth, was a trail of milk from the giants or gods; hence the name "Milky Way".

In fact, all fractals have the characteristic of self-similarity: any part is equal to the whole. We do not intend to make a course on fractals, since there are already articles and works on this subject. We simply wanted to offer this particularity that all fractals have.

The most usual representation of a turbulence is a spiral (everything depends on the parameterization that can be made of a phenomenon); it is as if the particles were attracted by a point (attractor), analogous to what happens in a sink, or as if the current lines are curled up on themselves around a point.

Every dynamic system has attractor points, understood as the points, states or times to which it is directed and evolves. If we parameterize a dynamic phenomenon in order to observe the current lines, the existence of a spiral (not zero rotational of the velocity field) will indicate the existence of turbulence. Under certain conditions, which are very easy to achieve, all dynamic phenomena governed by few simple laws, can cause turbulence; that is: under a potential context of chaos, it is easy to have turbulence or sudden alterations.

A few weeks ago, we found a series of photographs, specifically 4, in which the captions were wrong; the truth is that it was quite difficult for us to assign the captions to each one of them:

- Image from space taken to the vortexes created by the Island of Guadalupe.
- A Karman vortex channel created by a soccer ball with D=100.
- The distribution of galaxies within a cluster or group of galaxies and the distribution of stars within a given galaxy.

There are, and we are aware of this, many other phenomena, in principle different from all points of view, which are enormously similar in terms of their structure, representation or evolutionary dynamics. We find this structure or phenomenon, in situations so different, in scale, type or context, as: car wakes, sudden and chaotic alterations of the stock market or economy, turbulences in biological groups and information networks, internet, blocking of knots, politics, human relations, historical periods, medicine, psychology, meteorology, fluid dynamics, etc.

In short, turbulence is nothing more than "alterations or variations" with respect to, let's say, the "normal" of any type of dynamics; apparently, turbulence "twists" with respect to time (or to the chosen parameterization or representation); that is the essence of turbulence. This same "strange" or "abnormal" structure can also be found in other types of phenomena, not necessarily relative to the dynamics of a fluid.

An abrupt alteration of the stock exchange, for example, is nothing more than a turbulence caused by a series of initial and boundary conditions, which applied to a series of simple laws, originate a chaotic dynamic with respect to time. On countless occasions, after a bus has passed, we have related the turbulence it leaves behind to the results or alterations that certain economic policies, for example, leave in society.

The characteristic of self-similarity is shared by all dynamic phenomena, whatever their context.

Turbulent displacements are very common, both in nature (atmospheric flows, rivers,...) and in different applications of technological interest (flows in ducts, turbo machinery, boilers, combustion chambers, heat exchange equipment, vehicle aerodynamics,...), to the extent that most of the flows of interest, from almost any medium-serious or applicable point of view, are turbulent. The existence of turbulence alters various physical parameters of the fluid itself, as well as the flow itself.

For this reason, it is necessary to understand and comprehend its origin in order to be able to analyze and predict it.

The two scientific theories or methods could be said to be the most important and the most established, and par excellence, are radically different and propose extreme things:

- The existence of universal laws, which govern everything: a reductionist approach to science.

- The previous theory is not enough; to explain the world much more is required; from any level or scale, new phenomena appear, rich and varied, with elements absent in the previous, simpler level; new symmetries are generated and new forms of organization emerge; hence the need to generate theories for each phenomenon or even for each scale.

What is the real one? Perhaps, as is always the case in science and in knowledge in general, it is a mixture of booths.

Sir Horace Lamb (1849-1934), in an international tribute given to him on his eightieth birthday, in 1929, said: "When I die, I hope to go to heaven. There, I hope to be enlightened about the solution of two problems, quantum electrodynamics and turbulence. On the first, I am very optimistic...

The first was solved by Richard P. Feynman (1918-1988), for which he was awarded the Nobel Prize in 1965. Feynman: "turbulence is the last major unsolved problem in classical physics".

Numeric model for calculating fractals dimensions (in order to know if it is fractal-multifractal):

Haussdorf's definition of dimension was introduced in 1918.

Let "A" a set of "$\mathbb{R}n$"; the external measurement is defined α-dimensional of "A", as that: let "B_i" a coating of "A_i"; let "ε_i" the diameter of each coating and be an arbitrary epsilon.

$$B_A = \left\{ \{B_i\} \,/\, i = 1,...,\infty \,/\, A \subset \bigcup_i B_i \,/\, \varepsilon_i \leq \varepsilon \,/\, \forall i \right\}$$

We define external measurement α-dimensional of set "A", as that.

$$S_{\alpha,\varepsilon} = \lim_{\varepsilon \to 0} \inf_{B A} \sum_i \varepsilon_i^{\alpha}$$

For every "A" set, there is only one number "DH ", named dimension Haussdorf of "A", DH(A), for which the following is verified:

$$S\alpha(A) = \infty \quad \alpha < DH$$

$$S\alpha(A) = 0 \quad \alpha > DH$$

The topological dimension of a set is defined as the number of coordinates needed to express a point belonging to that set. We can remember that a fractal is that geometric structure whose Haussdorf dimension is strictly greater than its topological dimension; given a fractal structure, we obtain from it a unique Haussdorf dimension. There are several definitions of Haussdorf's dimension applied to certain geometry; the most used method to calculate the "DH" is the box-counting method; this value, although it does not coincide exactly with the Haussdorf's dimension, coincides in the most interesting cases. The method consists of the following:

It place on the figure to be studied, in 2 or 3 dimensions, a rectangular grid with an epsilon "ε" amplitude; I count the number of cells in which the figure enters some box and I call it "$N(\varepsilon)$". We repeat the experience for various values of reticular amplitude, and we place on a 2-dimensional graph, the values of "$\log(N(\varepsilon))$" and "$-\log(\varepsilon)$"; if we find the slope of the regression line that joins all the points found, we will have the dimension sought.

The problem arises when the set "A" to be studied is formed by points; it is true that we can also join these points by means of lines, and thus obtain a figure; but we do not intend to do this; we want to be able to study the discrete set of points, without having to generate another geometric figure from it. To do this, we need a series of concepts.

Let "A" be a set of points belonging to $\mathbb{R}n$. $A=\{Xi\}$ / i=1,...,n. Let's give this set "A", a measure of probability. We will count for each "Xi" the number of points of the set that we find inside a sphere of $\mathbb{R}n$ of radius "ε", with the center said "Xi". I will call "$ni\ (\varepsilon)$", to this value. The probability that we will associate to each point of "A", will be the following one: $pi\ (\varepsilon) = ni\ (\varepsilon)/N$. In order that it is really a measurement of probability, we will impose the following condition:

$$\sum_{i=1}^{N} p_i(\varepsilon) = 1$$

By Jensen et.al.1985, we have that the probabilities are related to the radius of the sphere, through a law of powers:

$$p_i \approx \varepsilon^{\alpha i}$$

The exponents "α_i", will be the characteristic exponents of the set "A". We can therefore consider the application that associates each point with its exponent: $V:Xi \rightarrow \alpha i$; studying how these values are distributed, will give us a lot of interesting information when it comes to characterizing set "A", with respect to dimensionality.   A multifractal, is a fractal that has a distribution of characteristic exponents, instead of a single exponent.

That is: there are different zones with different fractal structures.

For each "B_i" covering of the "A" set, of epsilon amplitude, we can consider the following partition function:

$$\Gamma_1(q, \tau, \{B_i\}, \varepsilon) = \sum_i \frac{P_i^q}{\varepsilon_i^{\tau}}$$

"P_i" is a measure of the "B_i" set. To eliminate the dependence with "B_i" we make the limit of "Γ_1" for "q" and "τ" less or equal to 0; if we call "Γ_2" to this limit, we can define the partition function, as follows:

$$\Gamma(q, \tau) = \lim_{\varepsilon \to 0} \Gamma_2(q, \tau, \varepsilon)$$

Define $\Gamma(q, \tau)$ = ∞   if $\tau > \tau(q)$

           =   0   if $\tau < \tau(q)$

Define function: δq: $\delta q = (q-1) - 1\tau(q)$

One of the reasons why this last definition is important, is $\delta 0 = DH$.

$\Gamma(0, \tau) = \lim (\inf \quad \varepsilon i - \tau) / \varepsilon \to 0 \quad \tau <= 0$

If it change "$\alpha = -\tau$", it have:

$$\Gamma(0, \tau) = \lim_{\varepsilon \to 0} \inf_{\tau \leq 0} \sum_i \varepsilon_i^{\alpha} = S_{\alpha}(A)$$

Since Haussdorf's dimension is unique, it follows that $\delta 0 = DH$.

The generalised dimensions of Rènyi are defined as follows: Grassberger, 1983:

$$\lim_{\varepsilon \to 0} \frac{- \sum\limits_{i=1}^{N} p_i(\varepsilon) \log(p_i(\varepsilon))}{\log(1/\varepsilon)} = D_q(\varepsilon)$$

$$D_q(\varepsilon) = \lim_{\varepsilon \to 0} (1-q)^{-1} \frac{\log \sum\limits_{i=1}^{N} (P_i(\varepsilon))^q}{\log(1/\varepsilon)}$$

These types of dimensions were introduced in 1983 by Hentschell and Procacia, as an attempt to define dimensions that were easier to establish relationships between them.

With the appearance of these dimensions, not only was this achieved, but also the dimensions used until then were defined, based on the generalized ones; dimension or capacity of Kolmogorov "DK" The following dimensions are available: the information dimension "DI" and the correlation dimension "DC".

$$DK = \lim_{q \to 0} Dq$$

$$DI = \lim_{q \to 1} Dq$$

$$DC = \lim_{q \to 2} Dq$$

The number of times "α" takes a value within the closed interval $(\alpha', \alpha' + d\alpha')$, can be expressed as follows:

$$n(\alpha')d\alpha' = \varepsilon^{-f(\alpha)} d\alpha'$$

"f(α)",is the fractal dimension of the subset of "A" that have the same characteristic exponent "α"; that is: "$\forall$ -1(α" where $\alpha \in$ (α' , α' + dα').

To draw the function "f(α)",we'll draw the envelope to the straights "y=qx-τ(q)", variating "q" from "-∞" to "+∞", (V.Martínez 1988).

The curve thus drawn has a single maximum and the value that the function takes at that point is the Haussdorf dimension of the set.

Very interesting things can be found in the works of P.Martien, S.Pope, P.L.Scott and R.S.Shaw on the time intervals between drops on a dripping tap in 1985, or the works of A.Provenzale, R.Vio and S.Cristiani on the variations of luminosity of quasar 3C-345 in 1993. From an apparently chaotic phenomenon in terms of its high unpredictability, we obtain a defined geometry, on which we define and calculate its multifractal dimensions. This characterization from the point of view of fractal geometry is applicable to other types of phenomena or time series.

The fractal dimension of a random distribution of galaxies is D = 3; the fractal dimension of galaxies located on the walls of low-pressure zones is D = 2; D = 1 on the walls, and D = 0, in galaxy clusters; in fact: D(r) with distance r (Mpc/h):

D(0) $\approx$ 1.5
D(0.8) $\approx$ 0
D(3) $\approx$ 2
D(100) $\approx$ 3

I have made an application – software in Mathcad, in order to apply it to dates (spatial coordinates, temporal series, etc....), and analyze the multifractal geometry characteristic (based in mathematic theory expose before):

$$h(n,j) := if\left[\left(x_j - px_n\right)^2 + \left(y_j - py_n\right)^2 < \varepsilon^2, 1, 0\right]$$

$$R_n := \sum_j h(n,j) - 1$$

$$pl_i := \frac{R_i}{N}$$

$$v := \sum_i pl_i \qquad P_i := \frac{pl_i}{v}$$

$$\alpha_i := \frac{log\left(P_i\right)}{log(\varepsilon)}$$

In this moment, it applies this procedure, in a Meteorites rain. That is: as a first sample:

It has a table with dates, each corresponding to the instant of sighting a Quadrantides (meteorites rain in 1992); the accuracy of the shots is the maximum achievable from the combination of the human eye, the brain, a paper and a pencil; the format of these shots is as follows:

- The first two decimal places correspond to the hour in universal time, the next two to the minutes and the last two to the seconds.

The table is composed of 303 time dates, and may be sufficient to obtain some information. For example: 0.015038.

In order to observe, analyze and draw some kind of conclusion about the moments of sighting,

it is convenient to normalize them by converting them into Julian time of day.

For this, assuming that in the data file, we have the moments already mentioned, and knowing that "a" is the year, "m" the month and "d" the day:

$$t :=$$

$$\text{C:\\..\\datos}$$

$$i := 2.. \, \text{length}(t) - 1$$

$$a_i := 1992 \qquad m_i := 1 \qquad d1_i := 3$$

$$fd1_i := \text{floor}\left(t_i \cdot 100\right) 60 \cdot 60$$

$$fd2_i := \text{floor}\left[\left[\left(t_i \cdot 100\right) - \text{floor}\left(t_i \cdot 100\right)\right] \cdot 100\right] \cdot 60$$

$$fd3_i := \text{floor}\left[\left(t_i \cdot 10000 - \text{floor}\left(t_i \cdot 10000\right)\right) 100\right]$$

$$fd4_i := fd1_i + fd2_i + fd3_i$$

$$d_i := fd4_i \cdot \frac{10}{24 \cdot 60 \cdot 60} + d1_i$$

d_j, is Julian day.

$$dj_i := \text{floor}\left(365.25 a_i\right) + \text{floor}\left[30 \cdot 6001 \cdot \left(m_i + 1\right)\right] + d_i +$$

179

$$+\left[2-\text{floor}\left(\frac{a_i}{100}\right)+\text{floor}\left(\left(\frac{\text{floor}\left(\frac{a_i}{100}\right)}{4}\right)\right)\right]$$

In this way, we can already work with this data, since it is already standardized and unified.

The geometric representation of the data we have is basic for a good analysis; and not only that: from certain geometry.

It will only be possible to extract certain information and not another one. It sees the different representations.

Let's see how the time intervals between each sighting evolve:

$$\text{dif1}_j := \text{dj}_j - \text{dj}_{j-1}$$

If it calculate the differences of a higher order and represent the evolution, it will obtain:

$$\text{dif2}_j := \text{dif1}_j - \text{dif1}_{j-1} \qquad \text{dif3}_j := \text{dif2}_j - \text{dif2}_{j-1}$$

,etc.

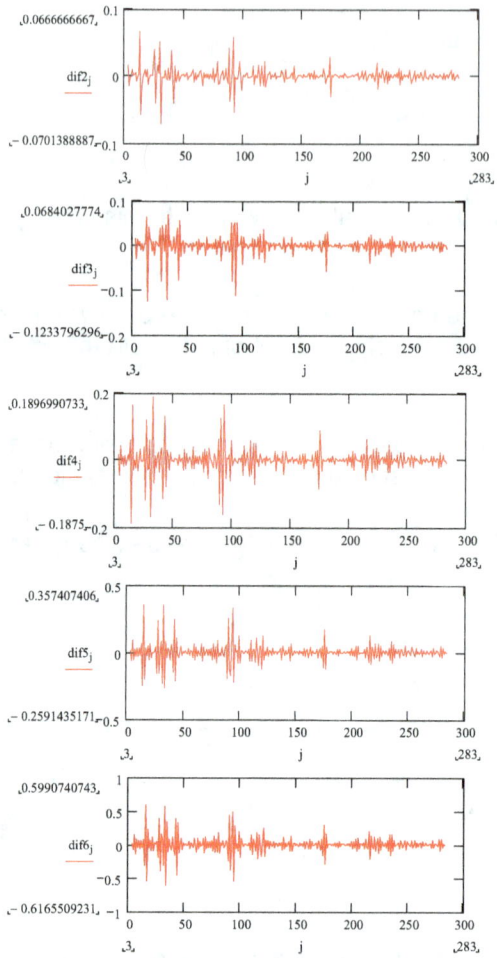

The areas with the largest peaks correspond to the time intervals of least activity and vice versa.

This representation does not tell us much, except for what has already been mentioned; let's take the following representation:

$$j := 3 .. \operatorname{length}(t) - 1$$

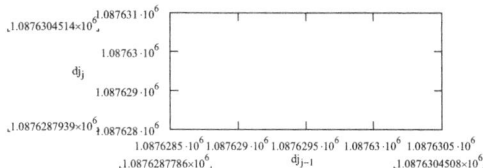

This geometry tells us the rate of growth of the time intervals between sightings; that is: as time goes by, the frequency of appearance of a Quadrantides decreases.

Let us now take the n-order differences:

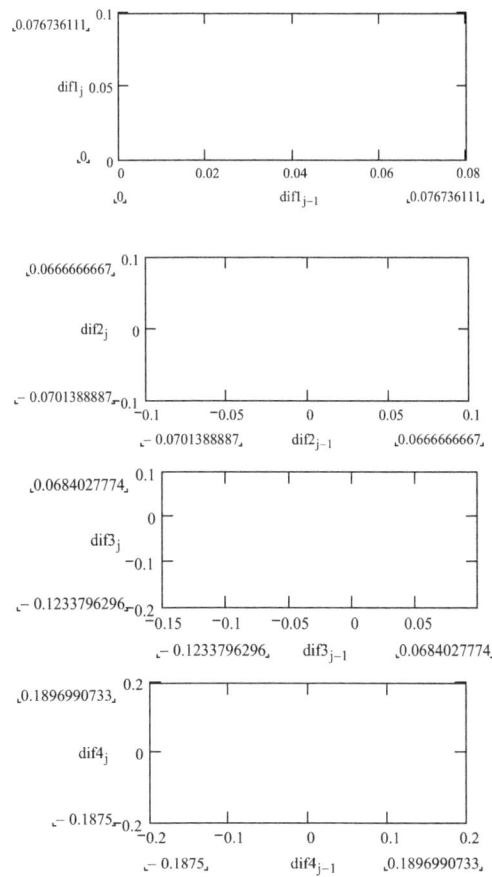

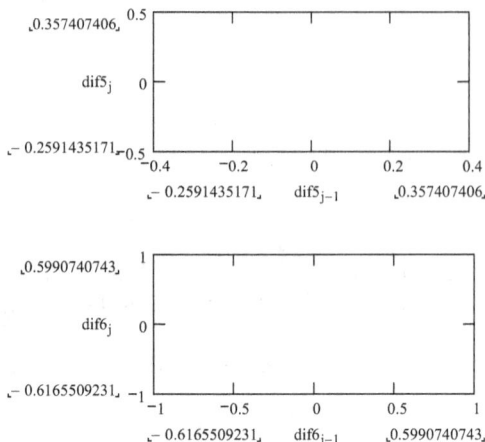

It has represented the differences of different orders, from consecutive values; we obtain analogous geometries, for non-consecutive values.

In this last representation-geometry, it applies the Multifractal theory.

Obviously, we could generate many more geometries, starting, for example, by substituting division, multiplication or whatever.

Another very useful representation is the three-dimensional one (even 4, etc....):

$$x(j) := dj_{j-1} \qquad y(j) := dj_j \qquad z(j) := dj_{j-2}$$

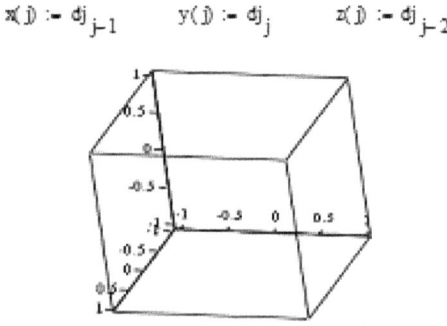

There is an Article ([38] Y. Brenier. U. Frisch. M. H´enon. G. Loeper. S. Matarrese. R. Mohayaee. A. Sobolevski):

From the actual Universe in high scale, trough "reverse engineering", is possible create the early structure and mass distribution.

The formation of filaments is completely logical and normal. If a group of mass particles starts with a constant density and all of them with the same mass, no filaments will be formed, but that (initial distribution "constant and uniform" it´s impossible (→ BMC – This BMC, it´s a initial instability ¡¡¡¡):

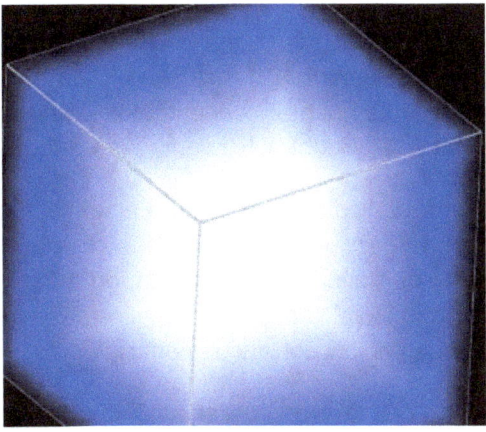

There's a particular structure, called Laniakea:

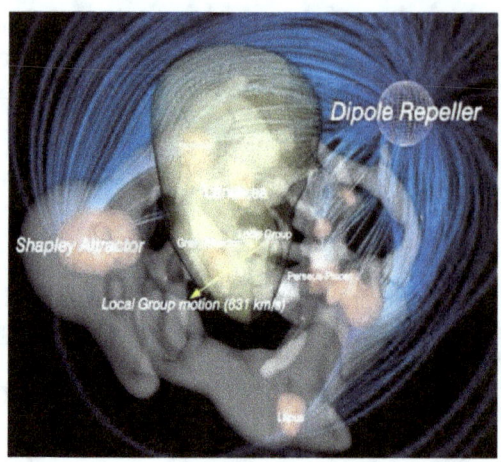

The yellow surface represents dust (latest data from Rosine Lallement and her team), and the purple surface represents hot star concentrations (and traces the outline of the Local Bubble). No dark matter though.

In the particular case of the Laniakea, "attractors" are observed that are accumulations of matter, which produce a lot of gravitational attraction; on the other hand, there are "repulsors" or dipole repeller, which are the opposite (zones of low density of matter).

Therefore, it is necessary to simulate gravity, to have a good numerical model of the evolution of the Universe on a large scale; "?" is a dipole repeller point or zone:

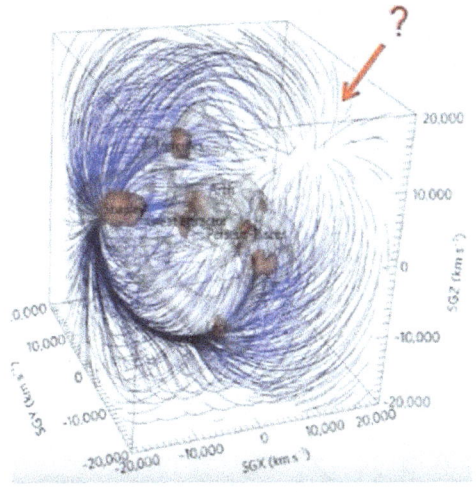

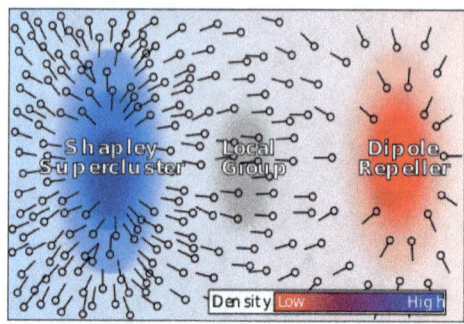

Density Low   High

The translation speed of galaxies today is not only explained by the existence of an attractor: a repulsive dipole is needed...:

The velocities of a galaxy can be calculated using gravitational fields. But we have to work with the possibility that the galaxies, in their paths, have been helped by the gravities of other galaxies or galaxy paths (gravitational aid). That is why the calculations, without working with this possibility, are incorrect.

It is necessary, therefore, that repellent dipole, but it is simply that it has less force of gravity or attraction, because it has less mass or density.

It is all a question of mass distribution and therefore of gravity: always attractive, never repulsive.

For example, in the case of a pipe, the fluid is expelled because there is a pressure difference: the fluid flow has a direction towards the low pressure. But this does not mean that at the other end, there is a "repeller" that pushes the fluid...:

ARTICLE 5

Particle dynamics numerical models

The dynamics of any particle evolves step by step; the particle that moves does so by moving at every instant; it does not know where its movement will end, but it decides at every instant, in which direction it wants to go.

If the reason for this decision is known, the complete dynamics is known; that is to say: it is enough to know the reason for one direction and not another; this is the essence of the discretization of equations and their analysis.

Any particle is subjected to a group of forces (pressure, magnetism, coriolis, etc.), the sum of which is the force with which it pushes the particle.

It is all a question of knowing where and with what acceleration each particle moves in its environment; that is: around the particle, depending on the size of discretization, there is the surface of a sphere; it is necessary to know a mathematical model that tells towards which point of that sphere the particle:

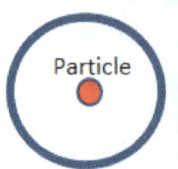 

It can define some numerical models to solve this problem:

-   *Brownian and DLA model*

If a set of particles possesses totally random displacements, at a certain moment the positions of these particles will be distributed randomly and uniformly throughout the whole space. That is Brownian movement.

In the case of not entirely random displacements, the final positions will not be uniform and uniformly distributed. In the reality, there is no phenomenon that has complete randomness, so the final distribution of a group of particles will not be uniform: threads, groups, etc. will always be formed.... randomly and uniformly throughout the whole space. This particle dynamic is called DLA or displacement of aggregates by limited diffusion (conditioned movement); for example: ice and lightning geometry formation:

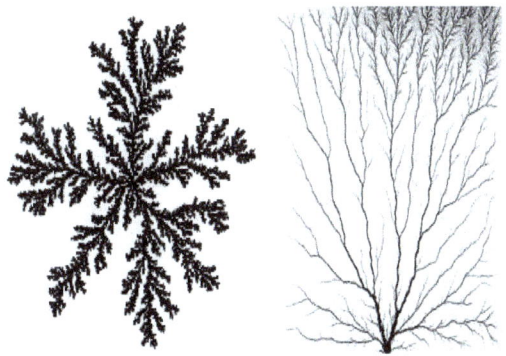

What is this factor-attraction between particles? The Viscosity, the gravity, etc. That is the same applied to humans and feelings. Also, a flock birds formation, is a particular case of DLA.

Brownian and not Brownian motion (very sensible to random motion limits); is practically impossible, to have a Brownian fully random; for example: initial particle point = (0,0) (Mathcad software):

$$x_1 := 0 \; y_1 := 0 \qquad i := 1.. \, 100000$$

$$alea1_i := if(rnd(1) < 0.5, -0.1, 0.1)$$

$$alea2_i := if(rnd(1) < 0.5, -0.1, 0.1)$$

$$x_{i+1} := x_i + alea1_i$$

$$y_{i+1} := y_i + alea2_i$$

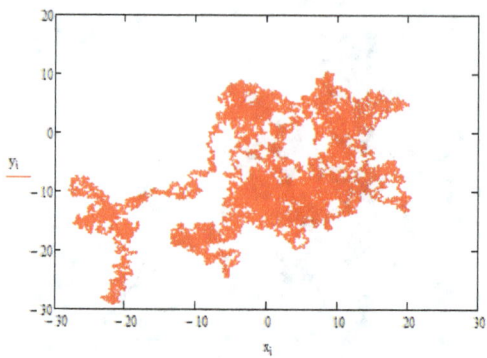

Brownian displacement conditioned (only 0.5%). Very good "filament"....:

$$x_1 := 0 \quad y_1 := 0 \qquad i := 1 .. 100000$$

$$alea1_i := if(rnd(1) < 0.505, -0.1, 0.1)$$

$$alea2_i := if(rnd(1) < 0.505, -0.1, 0.1)$$

$$x_{i+1} := x_i + alea1_i$$

$$y_{i+1} := y_i + alea2_i$$

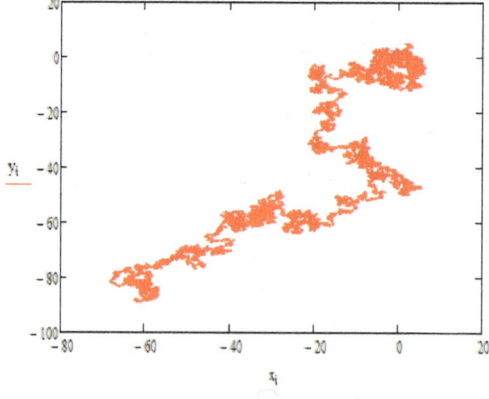

➔ The Brownian displacement "full random" is less likely: all movements, are DLA (with more or less conditionings).

This is the easiest method to decide: do not worry about the decisions of your environment.

Decide for yourself, with or without weights in some decisions. The path produced by this displacement, is a tube or zone of low pressure, so the matter, tend to aggregate there.

- *Navier Stokes equations*

The "traditional" Navier Stokes equations in general, are (without external forces; "g" or viscosity "μ" can be considered also as a "force" external and also Coriolis force and others may be):

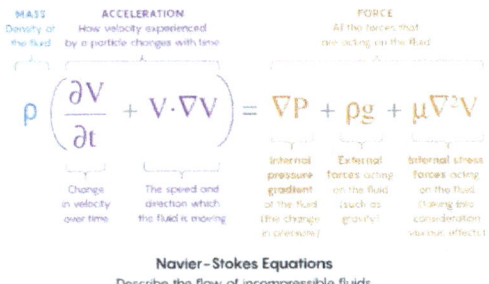

Navier-Stokes Equations
Describe the flow of incompressible fluids

One the first dynamic model from Navier Stokes equations, is: acceleration of suction:

$$\frac{\frac{\partial P}{\partial x}}{\rho} = \frac{P_x}{\rho}$$

This value or others terms of Navier Stokes, can be equated to acceleration, obtaining, different dynamic models:

$$\frac{\partial u}{\partial t} = -\frac{\frac{\partial P}{\partial x}}{\rho} - u\frac{\partial u}{\partial x} + F(viscous)$$

Obviously, in right terms, can adding the External forces as an electromagnetism, Coriolis effect, gravity, etc....

$$F(viscous) = \frac{\mu}{\rho}\frac{\partial^2 u}{\partial x^2}$$

- **Analogies between numerical models; analogies between some events**

It´s necessary to detect patterns in numeric models that describe the events, and then, it will be easier to detect patterns between events: it finds similarities between phenomenon's and numeric models.

It try to create a theory for explaining the distribution and evolution of matter in the Universe in large scale, galaxies dynamic, Universe expansion, Dark matter and dark energy, etc. But also is possible to apply this theory, in others fields as economy, human's relations, people flocks, stock market, feelings human, etc....

It tries to explain a general behavior in future, not a particular. All dynamic event in the cosmos, are a wave.... And as a wave, is necessary to study it.

Richard Feynman:

- "MATHEMATICS. To those who do not know mathematics it is difficult to get across a real feeling as to the beauty, the deepest beauty, of nature. If you want to learn about nature, to appreciate nature, it is necessary to understand the language that she speaks in."

- "A theory for a scientist, even it is your most desired wish, even if you have invested a lot of time, even if you have married, if not explain the reality, the theory is wrong."

Albert Einstein:

- "Look deep into nature, and then you will understand everything better."

A perfect and full description and analysis of Navier Stokes equations, is essential and necessary so.

Prey and depredator model

"x" number prey and "y" number predator:

$$\frac{dx}{dt} = ax - bxy$$

$$\frac{dy}{dt} = -cy + dxy$$

Is a model very simple with "x" and "y" initials and point fix (a/b, c/d):

$$\frac{dy}{dx} = -\frac{y}{x}\frac{dx - c}{by - a}$$

Phase space:

$$V = dx - c * \ln(x) + by - a * \ln(y)$$

Some images with different "x" and "y" initials, and "a", "b", "c" and "d":

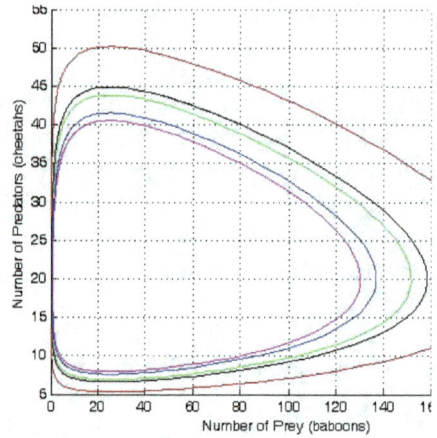

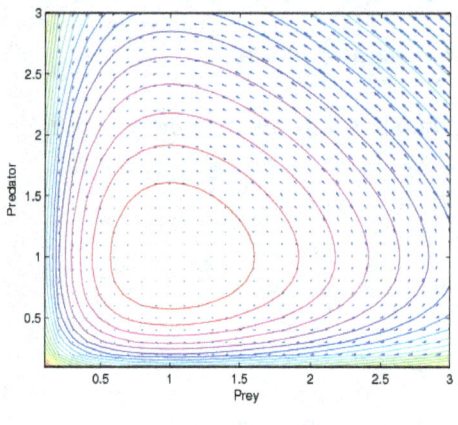

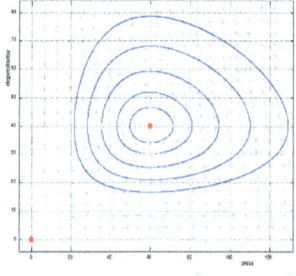

This geometry concept is very similar to: Typical Cavity problem fluids:

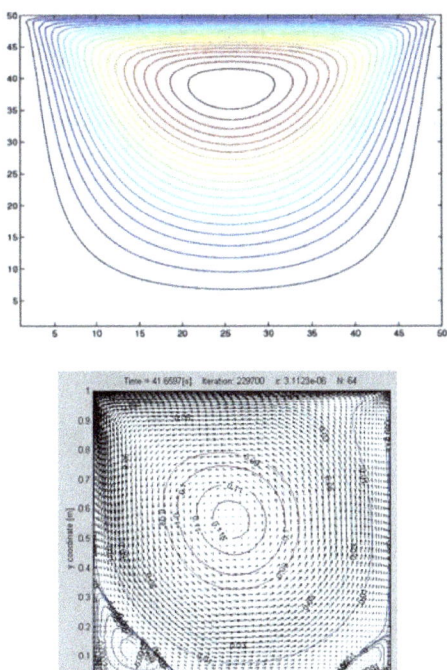

In the 2 last images, they are the representations of pressure lines in a cavity with a fluid in displacement.

Geometries very similar, numeric model, so, must to be also similar....

If it represents the solution of a typical problem in fluid dynamics such as the Cavity Problem, it obtains graphs like the following one, solving the Euler equations (or Navier Stokes):

$$u \frac{\partial u}{\partial x} + v \frac{\partial u}{\partial y} = -\frac{1}{\rho} \frac{\partial p}{\partial x}$$

$$u \frac{\partial v}{\partial x} + v \frac{\partial v}{\partial y} = -\frac{1}{\rho} \frac{\partial p}{\partial y}$$

The initial values and boundary conditions of this problem are:

- Velocities at each wall.
- Initial velocity.
- Of course, the fluid under study has a given density and viscosity.

The solution of this cavity problem, have a typical geometry showing the streamlines:

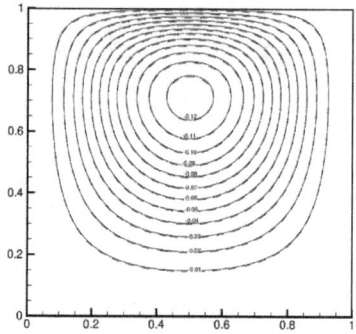

It is evident that there must be some way to solve the Prey-Predator problem with the Navier Stokes or Euler equations: must to be a relation between attractor point and wall velocities and other's dimensions and geometries; that is:

For example, in a cavity square of 1 length (Simulation CFD in Star ccm+ as a code):

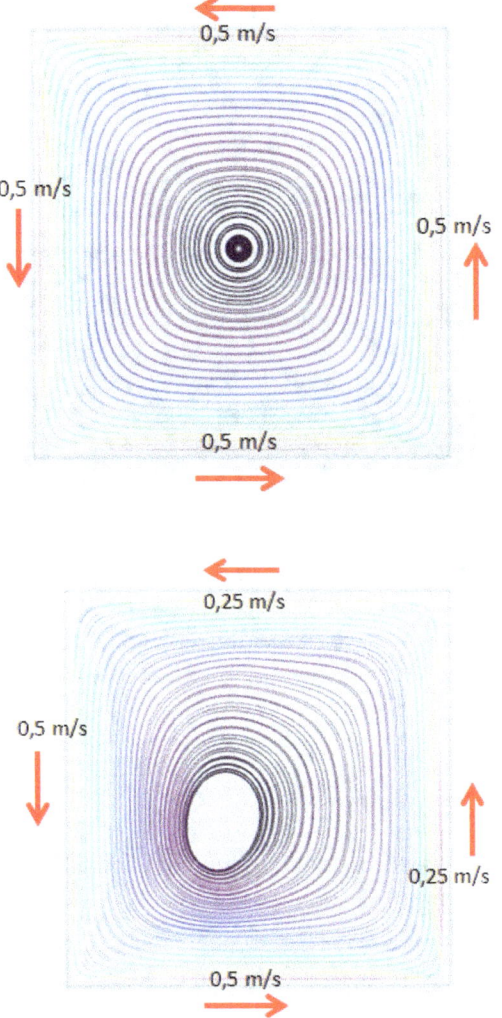

It´s not necessary, that the cavity to be square; may be circular or other geometry.

In these representation of space phases, is possible to change the orientation and scale, of axis. In the next image, we can see the displacement of a flock-group of sheep, in particular, in corner left down (and the zoom of this zone):

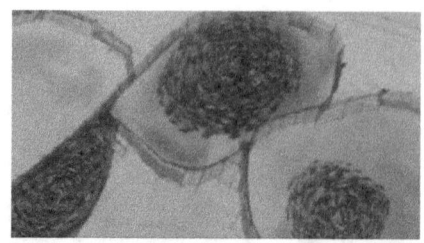

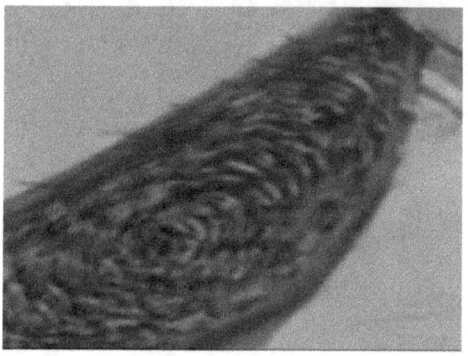

Is possible to know the vortex center in these models? Yes. When the variation of each variable (axis) is zero.

Romeo and Juliet model

The same happens in love equations between two peoples (Romeo "R" and Juliet "J" model):

$$\frac{dR(t)}{dt} = aR(t) + bJ(t)$$

$$\frac{dJ(t)}{dt} = cR(t) + dJ(t)$$

Second order derivatives can be added that specify functions that act as catalysts by accelerating or decelerating sentiment, such as economic stability, gender and family opposition, and include partial derivatives so that "R" and "J" do not depend only on "t".

"I love more a girl (H) if the girl (M) loves me":

$$\frac{dH(t)}{dt} = aM(t)$$

$$\frac{dM(t)}{dt} = -bH(t)$$

That is: the variation of my love to you, depend of your love to me.

There are other´s equations of love, one bit more complicate (Hannah Fry), but basically, are the same:

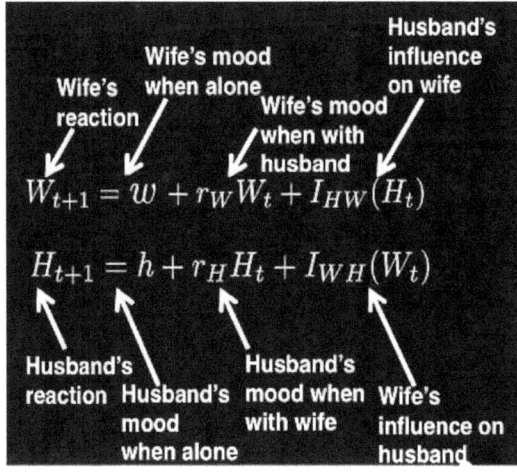

Lanchester model

In the Second World War, the Lanchester equations, for predicting an air combat ("A" and "B", number aircraft:

$$\frac{dA(t)}{dt} = -bB(t)$$

$$\frac{dB(t)}{dt} = -aA(t)$$

So, are the Prey and Lanchester equations, some similarities as a phenomenon? Are the Prey, Lanchester and Love, events similar? There are also, equations for war "guerrillas":

$$\frac{dA(t)}{dt} = -bA(t)B(t)$$

$$\frac{dB(t)}{dt} = -aA(t)B(t)$$

If the phenomenon is the "same", the numeric model also, but vice versa, is not necessary....

- In the Lanchester case eat aircrafts, and in the Prey case, eat animals, and if one go up, the other go down, with a gap or delay time.

- Basically, prey model and Lanchester model, are the same. It can transform:

$$ax - bxy \rightarrow x(a - by)$$

Black-Scholes model

It´s a model for analyze the behavior of Stock Market (sell and call), and predict some prices in the future.

The expression is very similar to Navier Stokes equations (parts):

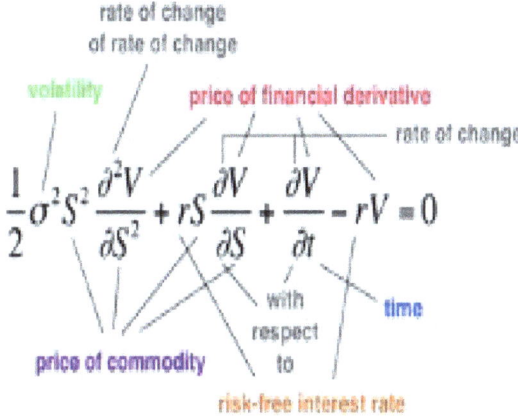

Other expression for Navier Stokes equations:

$$\frac{\partial u}{\partial t} + u\frac{\partial u}{\partial x} - fv = -\frac{1}{\rho}\frac{\partial p}{\partial x} + K\frac{\partial^2 u}{\partial x^2}$$

And its similarity with Black-Scholes:

$$1. \quad \frac{\partial V}{\partial t} \sim \frac{\partial u}{\partial t}$$

$$2. \quad \frac{1}{2}\sigma^2 S^2 \frac{\partial^2 V}{\partial S^2} \sim K\frac{\partial^2 u}{\partial x^2}$$

$$3. \quad rS\frac{\partial V}{\partial S} \sim u\frac{\partial u}{\partial x}$$

$$4. \quad -rV \sim -fv + \frac{1}{\rho}\frac{\partial p}{\partial x}$$

Schrodinger equation-model

$$i\hbar\frac{\partial}{\partial t}\psi(\mathbf{r},t) = -\frac{\hbar^2}{2m}\nabla^2\psi(\mathbf{r},t) + V(\mathbf{r},t)\psi(\mathbf{r},t)$$

Similar to Navier Stokes equations (parts).

It can work so considerer as a wave (there is a wave expression in Schrodinger equation), all event in the universe.

$$\Psi(x,t) = Ae^{i(kx-\omega t)} = A[\cos(kx-\omega t) + i\sin(kx-\omega t)]$$

Alan Turing biology evolution model

It´s a numeric model in order to predict the formation of patterns in the Nature (there is a wave expression in equations model):

$$\frac{\partial u}{\partial t} = D_u \frac{\partial^2 u}{\partial x^2} + f(u, v)$$

$$\frac{\partial v}{\partial t} = D_v \frac{\partial^2 v}{\partial x^2} + g(u, v)$$

"u" and "v" are the concentrations of 2 axis. "D_u" and "D_v" are the coefficients of diffusion of "u" and "v", and "f" and "g" the reaction between.

It can see perfectly, the heat equation (diffusion), into Alan Turing equations.

We can see also this evolution is Voronoi schemes (the evolution of "point", may be a reaction in function of time and also metric-distance):

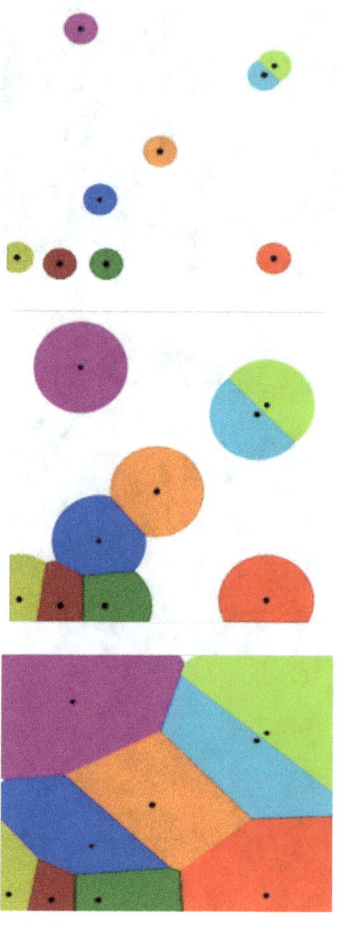

Other Voronoi generation schemes version: useful for meshing in CFD techniques:

"In computer science and electrical engineering, Lloyd's algorithm, also known as Voronoi iteration or relaxation, is an algorithm named after Stuart P. Lloyd for finding evenly spaced sets of points in subsets of Euclidean spaces and partitions of these subsets into well-shaped and uniformly sized convex cells. Like the closely related k-means clustering algorithm, it repeatedly finds the centroid of each set in the partition and then re-partitions the input according to which of these centroids is closest. In this setting, the mean operation is an integral over a region of space, and the nearest centroid operation results in Voronoi diagrams.

Although the algorithm may be applied most directly to the Euclidean plane, similar algorithms may also be applied to higher-dimensional spaces or to spaces with other non-Euclidean metrics.

Lloyd's algorithm can be used to construct close approximations to centroidal Voronoi tessellations of the input, which can be used for quantization, dithering, and stippling. Other applications of Lloyd's algorithm include smoothing of triangle meshes in the finite element method."

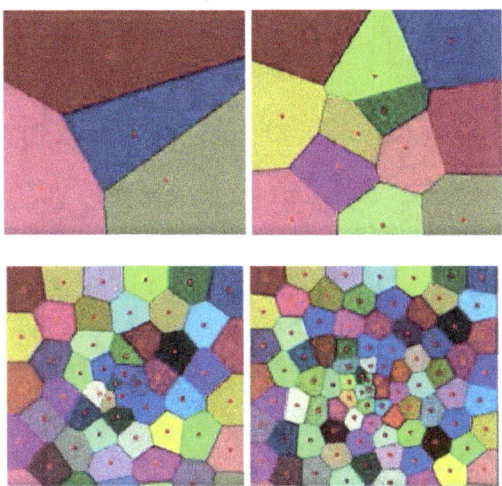

Heat equation

"T" is the temperature, "V" the velocity vector, "a" the acceleration vector and "x" and "y", the coordinates in 2D:

$$\vec{V} = (u, v)$$

$$\vec{a} = \frac{D\vec{V}}{Dt} = \frac{\partial \vec{V}}{\partial t} + \frac{\partial x}{\partial t}\frac{\partial \vec{V}}{\partial x} + \frac{\partial y}{\partial t}\frac{\partial \vec{V}}{\partial y}$$

$$\frac{\partial T}{\partial t} + U\frac{\partial T}{\partial x} + V\frac{\partial T}{\partial y} = a\frac{\partial^2 T}{\partial y^2}$$

➜ Considerations about:

In these models before, the numerical models are very similar, so the phenomenon must to be also (may be....).

??¿¿ Schrödinger, Black-Scholes, Alan Turing, ...:

In these 3 equations, it can see the diffusion equation (heat equation).

This diffusion part also is in Navier Stokes equations. If in Navier Stokes equations, the extern forces are zero, is possible create and apply the Alan Turing model.

➜ It´s possible so, apply Schrödinger equation, to Stock Market evolution.... ??

- *Sample 1: Navier Stokes in two dynamic events: Pedestrians and birds*

The model simplest for pedestrian dynamics, it´s witch the movement is caused by the pressure difference (suction) ("x" dimension, "m" mass, "P" pressure, "ρ" density, Kinetic and Potential energy):

$$KineticE = \frac{1}{2}m \overset{*\,2}{x}$$

$$PotentialE = m\frac{P_x}{\rho}x$$

So: $$P_x = \frac{\partial P}{\partial x} \overset{**}{=} x\rho$$

Other model: Navier Stokes with Viscosity as a gap time:

$$\frac{\partial \vec{V}}{\partial t} + \left(\vec{V}\nabla\right)\vec{V} = -\frac{1}{\vec{\rho}}\nabla\vec{P} + \frac{\vec{\mu}}{\vec{\rho}}\nabla^2\vec{V} + ...$$

It's possible to calculate the pressure "P" (in order to know, what this pressure is) (variation) from 1 real test; from this test, it know the density and the acceleration.

One example of the design of evacuation of pedestrians or people from an enclosure (study made by Arcadia.org, with team members: Mrk Clayton, Wei Yan): Given a corridor where people walk, at the end of which, there is an opening or a door.

In the first image, you can see the accumulation at the exit increasing (in red) the pressure or density. By placing small circles along the corridor, the density is reduced:

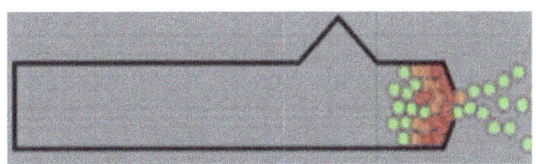

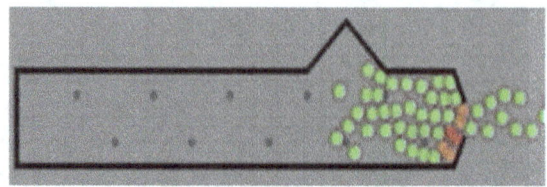

In the example I have made, you can observe exactly the same, before and after the placement of "obstacles", with the same objective: Reduce pressure or density and improve the flow rate in outlet:

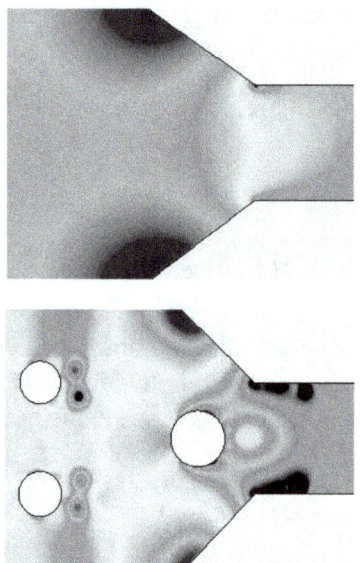

The "dark" zones, show the zones with low pressure or density, attracting peoples....

In Article ([39] Yongxiang Zhao, Meifang Li, Xin Lu and Ting Li), it analyzes the obstacles in "ducts" in order to reduce the "high pressure":

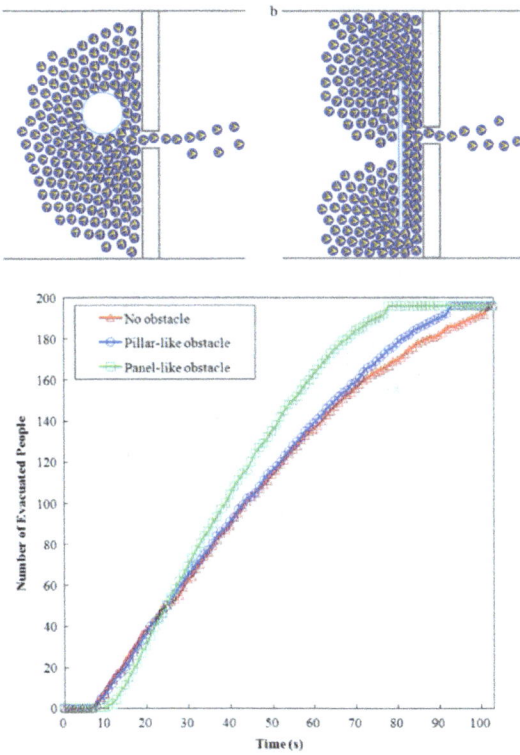

It is also possible to simulate the movement of pedestrians, by assuming that each person is a particle with certain characteristics of compressibility, comfort area, etc:

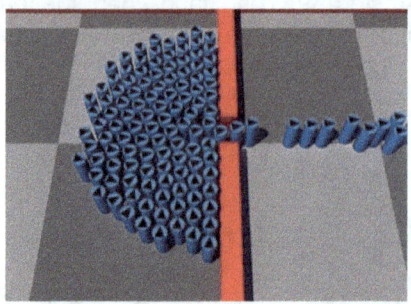

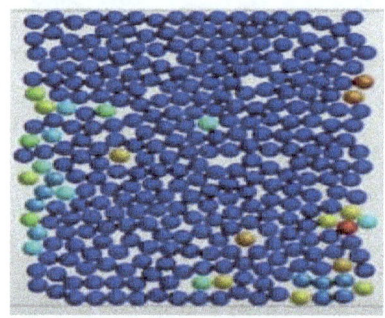

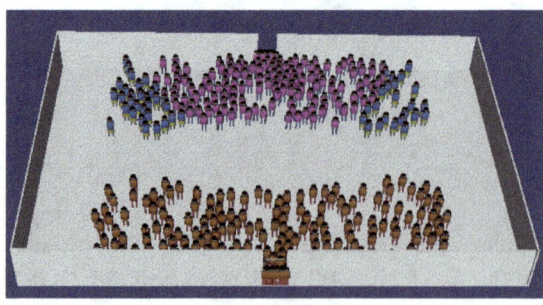

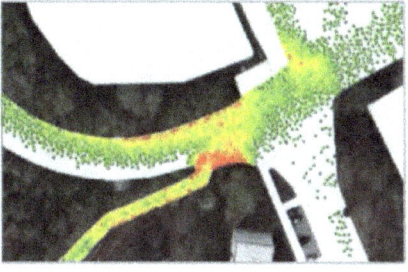

In fact, these models can be applied in other´s samples or fields as a fluids, traffic, birds, etc.…

a) Case 1

The simplest case of Navier Stokes equations is applied to the movement of pedestrians; it is a matter of applying the movement towards the minimum pressure.

The potential "U" of this dynamic is defined as the pressure variation, divided by density; this value is a potential, i.e. the maximum energy value that can potentially reach a particle (1 dimension or direction "x"):

$$U = \frac{\partial P / \partial x}{\rho}$$

Therefore, the action "S", defined as ("T" kinetic energy):

$$S = \int_{t1}^{t2} (T - U)dt\,)$$

In this case particular ("V" velocity):

$$T = V^2$$

For the action to be minimal, the potential has to be maximum; this coincides with previously explained that the particle will go towards the maximum pressure variation (divided by density) possible.

Note: remember that "V", "ρ", "P" are vectors, so "T" and "U" also.

It´s possible in this simple and first case, apply the Navier Stokes equations, adding more terms as externals forces.

b) Case 2

This numerical model can be improved by adding to the potential, the friction energy that opposes the movement of the particle; that is: the Viscosity, that depend of the density:

$$U = \frac{\partial P / \partial x}{\rho} + ...$$

From the Brownian displacement in which the particles move without restriction, passing through aggregates of limited diffusion (DLA) where there is only one restriction or condition of displacement, following the displacement of planets, all the dynamics obey rules of displacement between the particles that they make up the group, extremely simple and easy: flocks of birds or pedestrians, are clear examples of this fact:

By Craig Reynolds:

"" In 1986, I made a computer model of coordinated animal motion such as bird flocks and fish schools.

It was based on three dimensional computational geometry of the sort normally used in computer animation or computer aided design. I called the generic simulated flocking creature's boids. The basic flocking model consists of three simple steering behaviors which describe how an individual boid maneuvers based on the positions and velocities its nearby flock mates:

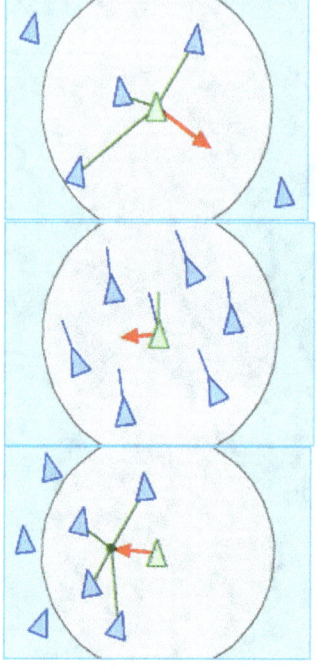

Separation: steer to avoid crowding local flock mates.

Alignment: steer towards the average heading of local flock mates.

Cohesion: steer to move toward the average position of local flock mates.

Each boid has direct access to the whole scene's geometric description, but flocking requires that it reacts only to flock mates within a certain small neighborhood around itself. The neighborhood is characterized by a distance (measured from the center of the boid) and an angle, measured from the boid's direction of flight. Flock mates outside this local neighborhood are ignored. The neighborhood could be considered a model of limited perception (as by fish in murky water) but it is probably more correct to think of it as defining the region in which flock mates influence a boids steering:

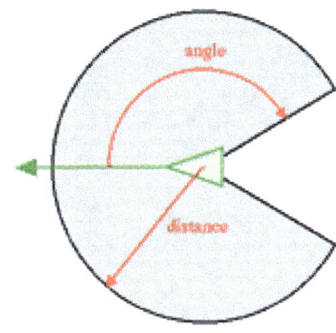

A boid's neighborhood.""

There are lot theories about this special dynamics phenomenon. For example, other theory or numerical model:

The path of one particle is the path with the minimum pressure. Also, the pressure work as a density; one particle will be where there is less density. So is possible create algorithms in order to create the path for any particle. Give a particle and give "sectors" in a sphere with center the particle (only a part in Navier Stokes equations) ("u" velocity in the coordinate (working in 1 dimension); "n" means the position "n" in direction about):

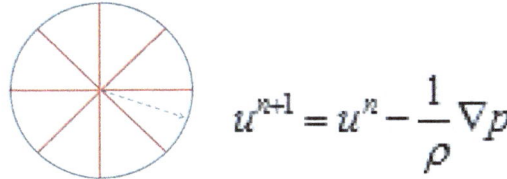

$$u^{n+1} = u^n - \frac{1}{\rho}\nabla p$$

The particle will move toward the half angle line of sector, with the least density (pressure); this displacement, with a delay time (viscosity) (step by step).

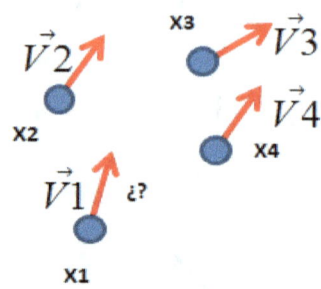

$$\vec{V1} = \frac{\vec{V2} + \vec{V3} + \vec{V4}}{3}$$

May be work with the same laws for pedestrians? may be.... It should be necessary, add some terms....

- *Minimum energy-length model*

Any particle, tend to walk with the minimum energy; Lagrangian "L", is defined as Kinetic Energy minus Potential Energy. The Potential Energy, indicates the reserve energy or the energy that the particle still has to be used:

$$E = E_K - E_P = T - U = L$$

To calculate by a Lagrangian or Classical Mechanical method, the path of a any particle, can be done:

- Discreetly or punctually.
- Continuously.

In the first case, given a particle moving in an instant, the position will be calculated in the subsequent instant, applying the minimization of the Lagrangian "L"; in the particular case "Newtonian" (non with Relativity), about a particle in orbit with "n" masses: ("V" velocity, "M" and "m" masses, "G" constant gravitation, "r" vector between the 2 masses):

$$\frac{1}{2}mV^2 - \sum_{i=1}^{n}\frac{GM\,m_i}{r} \to mimimum$$

In the second case, the Euler Lagrange Equations are developed, to know the equations that describe the dynamics of the particle (conservative forces; not dampers for example):

$$\frac{d}{dt}\left(\frac{\partial L}{\partial \overset{*}{x_i}}\right) = \frac{d}{dt}\left(\frac{\partial L}{\partial x_i / \partial t}\right) =$$

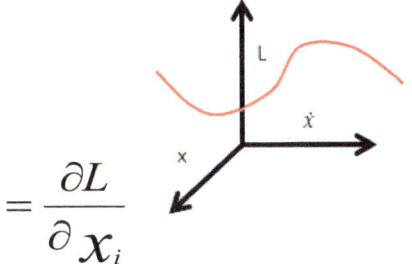

$$= \frac{\partial L}{\partial x_i}$$

The "paths" ideal, the correct path, ideal or real, is one for which the following integral is minimal:

$$\int_{a}^{b} L\,dt$$

From this concept of minimum energy to follow path, it is established that it is good:

- To have a lot of POTENTIAL "V" energy.
- To have little KINETIC "T" energy.

This, it´s the "good" direction in a path….

That is:
The path always follows a minimum distance path (geodesy) between 2 points, in a given metric, in a given space or surface.

A particle follows a path towards a state of minimum energy; of all possible paths, choose the one with a minimum Action ("S"); the difference between kinetic energy "T" and potential "U" is called Lagrangian "L"; this is a basic principle ("t1" the initial time of the beginning of the movement and "t2" the final instant); "T" and "U" will change, depending on which metric or space the particle is in. If it discrete the time and space, it can tell where a particle will go at any instant; this is exactly what is needed:

$$S = \int_{t1}^{t2} (T - U)\,dt$$

For example: what height will a particle reach with an initial velocity, on an inclined plane?

It is an easy problem to solve; but it is the essence of the calculation of geodesies and therefore, of any trajectory; how? (river trajectory, for example):

The possible directions that the particle can take are discretized; the differences between kinetic and potential energy of each possible path are calculated; the path that has the smallest of these subtractions will be the path chosen by the particle.

Some samples about Euler Lagrange Equation:

Application about pendulum dynamic:

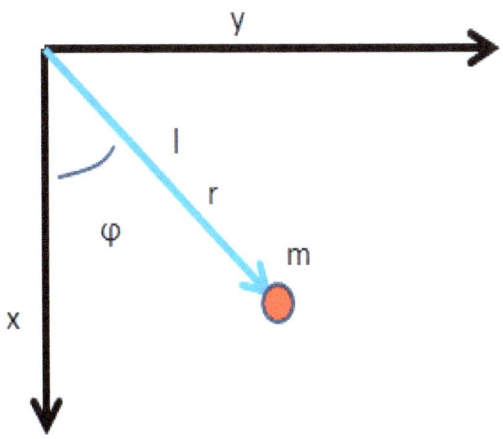

$$L = T - V = \frac{1}{2} m v^2 + mgx$$

$$\vec{r} = \left(l \cos \varphi, l \sin \gamma \right)$$

$$\overset{*}{\vec{r}} = \left(-\overset{*}{\varphi} l \sin \varphi, \overset{*}{\varphi} l \cos \varphi \right) = v$$

$$x = l \cos \varphi$$

Applying Euler-Lagrange:

$$\overset{**}{\varphi} + \frac{g}{l}\sin\varphi = 0$$

Application about down fall body:

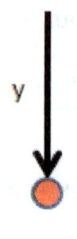

y

$$L = \frac{1}{2}m\,\overset{*}{y}{}^{2} - mgy$$

$$\frac{\partial L}{\partial y} = -mg$$

$$\frac{\partial L}{\partial \overset{*}{y}} = m\,\overset{*}{y}$$

$$\frac{d}{dt}(m\,\overset{*}{y}) = m\,\overset{**}{y}$$

$$F = -mg$$

Normally or at least, with velocity not big, the metric is the Euclidean. But there are more cases, for example with relativity or high mass and velocity; it's necessary other metric (distance definition):

Schwarzschlid metric:

$$ds^2 = -\left(1 - \frac{2GM}{c^2 r}\right)cdt^2 + \left(1 - \frac{2GM}{c^2 r}\right)^{-1}dr^2 + r^2\left(d\theta^2 + sen^2\theta d\phi^2\right)$$

In these equations, is possible to substituting in "r", the distance in every space-metric, in order to calculate

geodesics (minimum distance) (changing also "m", etc....):
for example: distances for Taxis in Manhattan: distance or
metric, is equal to "d" or "Manhattan metric":

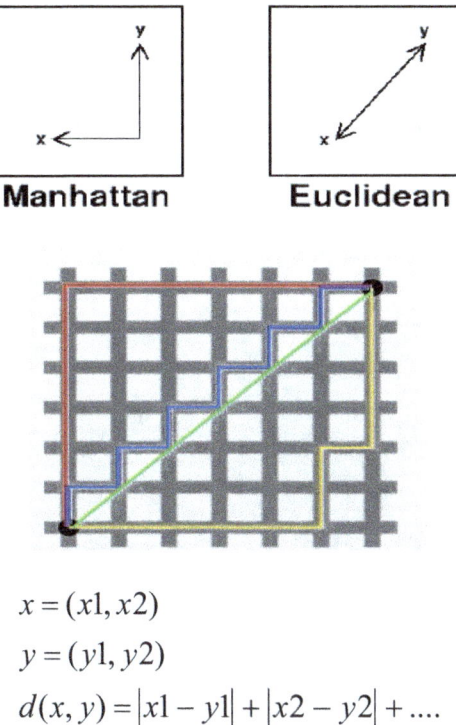

Manhattan   **Euclidean**

$$x = (x1, x2)$$
$$y = (y1, y2)$$
$$d(x, y) = |x1 - y1| + |x2 - y2| + ...$$

With this metric, a sample of the Voronoi mesh is:

But, the gravitation action (potential energy) is not an action with velocity infinite: the speed of gravitation is the speed of light (this force, change depending of density for example). This value is the "gap time action" (viscosity).

In this Potential Energy, in order to have more accuracy (real so), is necessary to add the Viscosity force (as a magnetic force or similar) and other´s forces.

One sample about geodesic calculation: it needs a geodesic in cylinder shape:

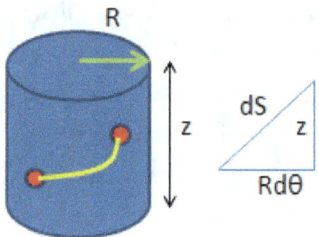

$$d\,S^2 = d\,x^2 + d\,y^2 + d\,z^2$$

$$x = R\cos\theta\,/\,y = R\sin\theta\,/\,z = z$$

$$S = \int_a^b L\,d\theta = \int_a^b \sqrt{R^2 + \overset{*}{z}}\,d\theta$$

$$\frac{\partial L}{\partial z} - \frac{d}{dx}\frac{\partial L}{\partial \overset{*}{z}} = 0$$

ARTICLE 6

Expansion of Universe models

Abstract

There are several numerical models, to explain how is the expansion dynamics, like the equations of Friedmann-Lemaître-Robertson-Walker.

But it is a model that is very simple to explain, for example, the variation of pressure or density, and other important values, and more: is only possible an acceleration negative. In this article new concepts and new procedures are defined, to obtain a more useful numerical models that are able to describe the dynamics of the Expansion of the Universe (positive or negative), as well as the evolution of diverse variables that participate in the phenomenon, as a variation density and pressure, and even the acceleration.

In these new created numerical models, expressions are established to calculate certain values of the intergalactic medium (such as density, viscosity, pressure, force), considering it as a fluid, which will be very useful in later articles, to know the evolution and interaction of matter in the Universe And other think more important: is possible other vision about Dark Energy, as force, not as a matter or particle. It describe expansion Universe model as a spring damper set, as energy from vacuum or low pressure and also and finally, from Navier Stokes equations.

Introduction

If it looks at the distances that separate us from the galaxies that are not in our local group, you will see that they are moving away from us. That is to say: the Universe is expanding.

Edwin Hubble already validated this fact experimentally by measuring the distances and speeds of many galaxies, and created a numerical model that expressed the linearity between distance "x" and speed "V", whose constant is called the Hubble constant "H" ($V=Hx$). Recent observations have shown that this relationship is not linear, and some more important: the Universe is expanding with acceleration ($H=H(t)$). Why is it expanding and how? This expansion, will be end?

They are 3 very important questions that currently, do not have totally satisfactory or real answer. All explanations are hypothesis, which as such, must be demonstrated, and explain the reality.

One of the most accepted but not validate, is that there is a kind of dark energy, which works as the opposite of gravity, that is: repelling instead of attracting. In fact, the evolution of length parameter ("a") in axis vertical, from big bang, against universe time in axis horizontal, from now: it´s possible see that from end inflation period to more or less 7500 million years, there are an acceleration negative, because the density was bigger than expansion force. From this date to actuality, surprise and Nobel prize so, there are acceleration positive (the gravity force was less than "dark force-energy"):

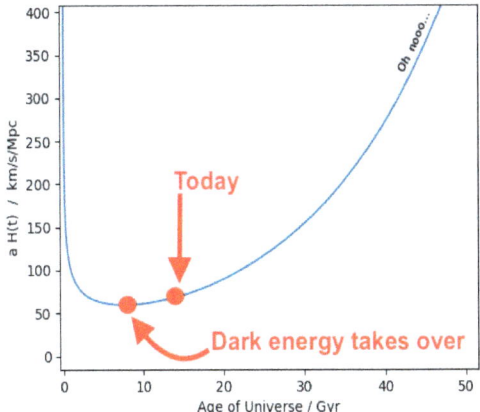

"H" is not constant in Time:

"a" scale universe factor (traditional nomenclature):

$$H(a) = H_0 \sqrt{\Omega_{R,0}\, a^{-4} + \Omega_{M,0}\, a^{-3} + \Omega_{K,0}\, a^{-2} + \Omega_{\Lambda,0}}$$

$$t(a) = \frac{1}{H_0} \int_0^a \frac{a'\, da'}{\sqrt{\Omega_{R,0} + \Omega_{M,0}\, a' + \Omega_{K,0}\, a'^2 + \Omega_{\Lambda,0}\, a'^4}}$$

Depending these values, it have one or other universe evolution, so is necessary know its….:

$$H_0 = 67.3 \text{ km s}^{-1} \text{Mpc}^{-1},$$
$$\Omega_{R,0} = 9.24 \times 10^{-5},$$
$$\Omega_{M,0} = 0.315,$$
$$\Omega_{\Lambda,0} = 0.685,$$
$$\Omega_{K,0} = 0$$

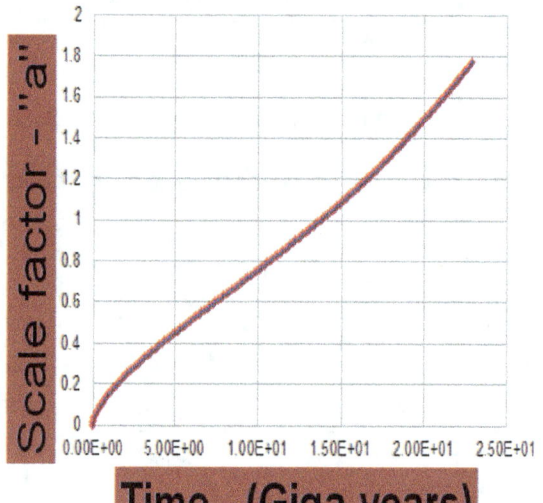

If $\Omega_{M,0} = 0.01$:

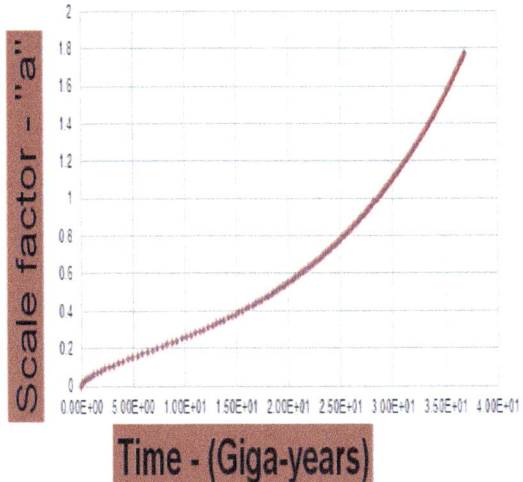

It's possible interpolate these lines, in this case, first sample and values:

$$a(t) \approx c_1 t^{2/3} + c_2 \left(e^{t/c_3} - 1 \right)$$

$$c_1 = 0.822$$
$$c_2 = 0.0623$$
$$c_3 = 0.645$$

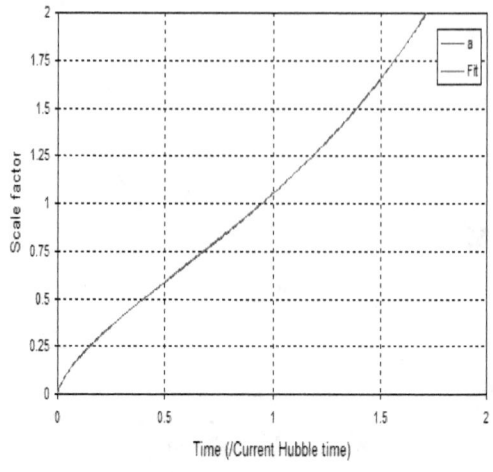

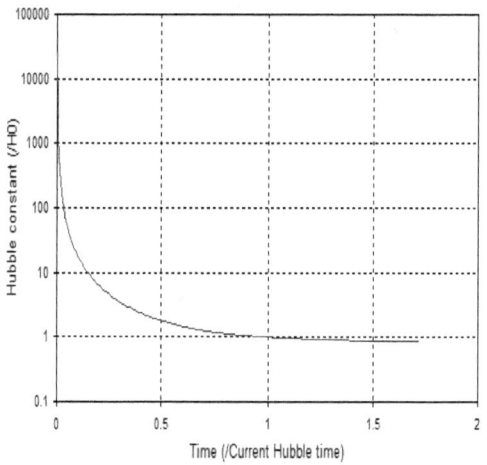

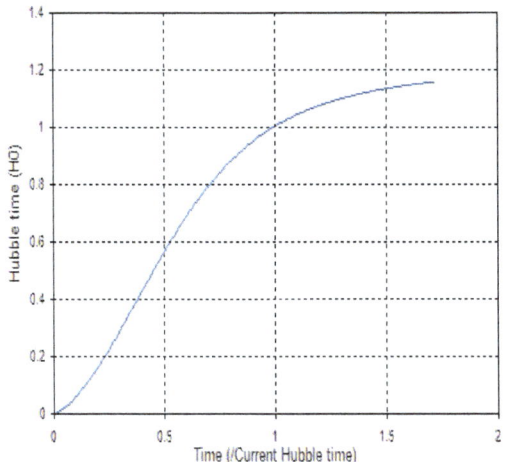

How to do the Excel sheet?

	A	B
1	Scale Factor Calculation	
2		
3		
4	Constants	
5	Omega R	9.24E-05
6	Omega M	0.01
7	Omega K	0
8	Omega D	0.99
9	H0	2.18E-18
10	Secs/gyr	3.16E+16
11		

	a'	f(a')	int(f(a'))	t		TIME	SCALE FACTOR
21							
22						t	a
23	0	0	0	0.00E+00		0.00E+00	0
24	0.01	0.720918704	0.003604594	1.65E+15		5.23E-02	0.01
25	0.02	1.169293969	0.013055657	5.99E+15		1.90E-01	0.02
26	0.03	1.512911245	0.026486983	1.21E+16		3.64E-01	0.03
27	0.04	1.797985443	0.043021166	1.97E+16		6.25E-01	0.04
28	0.05	2.043648414	0.062229336	2.85E+16		9.03E-01	0.05

0 =A23/(SQRT($B$5 + $B$6*A23 + $B$7*A23*A23 + $B$8*A23*A23*A23*A23))

0.003604594 =C23+(A24-A23)*(B24+B23)/2

0.00E+00 =C23/$B$9

0.00E+00 =D23/$B$10

"ρm" matter density, "ρR" radiation density, "ρΛ" dark energy density:

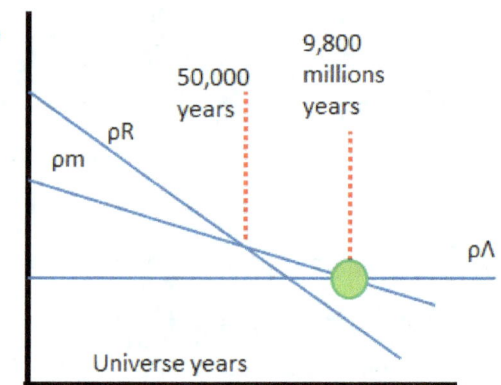

In green point, the matter attraction, win to dark energy; so the universe it expands ("c" speed of light, "P" pressure and "ρ" density):

$$P = w\rho c^2$$

"a" it´s a expansion factor (2 for example, means that the universe is 2 times bigger):

$$w = 0 \rightarrow barionic - matter$$
$$w = 1/3 \rightarrow radiation - matter$$
$$w = -1 \rightarrow dark - energy$$

$$\rho(t) = \rho m_0 a^3 + \rho R\, a^4 + \rho\Lambda$$

Hypothesis

- Hypothesis 1: Non Homogeneous Universe.
- Hypothesis 2: Non Isotropic Universe.

Velocity against distance expansion model

In city traffic, when the speed of cars is bigger, the separation between the, also in bigger:

Today, we know that in Hubble expression H=H(t); the expansion of Universe depend of a lot of things.

The "Expansion Velocity" may be, depend of Pressure and Density also ("x" space length), ("a" and "b" > 0):

$$V_E = function(x, P, \rho)$$

$$V_E \propto \frac{1}{\rho^a P^b}$$

$$V_E = H(t)\frac{1}{\rho(x)^a P(x)^b}x$$

And more: is possible, a little variation of Hubble law:

$$V = H x^{1+\varepsilon} / \varepsilon \neq 0$$

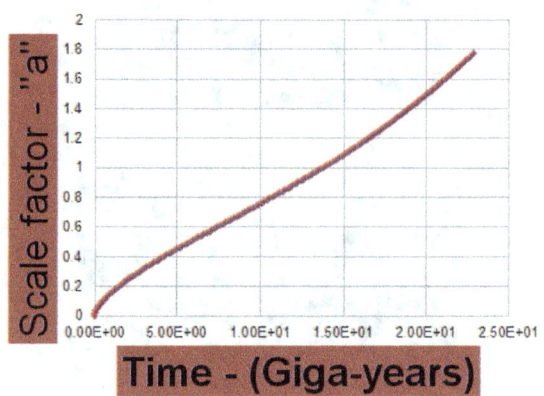

Time - (Giga-years)

Analyzing the possible error with "ε":

If $\varepsilon=0.01$, then (10^{26} meters→radio universe):

$$if \rightarrow x = 10^{26} meters$$

$$x^{2\varepsilon} \approx 1.8(error)$$

- **Vacuum acceleration expansion universe model**

Between 2 zones with different pressure, there is pressure difference which produces acceleration "a" (high to low pressure): pulling with acceleration "a" ("ρ" density, "P" pressure and "x" coordinate length):

$$a = \frac{1}{\rho}\frac{\partial P}{\partial x}$$

What is the origin of this Acceleration – Suction, as a Dark Energy action? It know perfectly, that the Bing Bang, is not an explosion or blast (is a space expansion). But, is perfectly possible to assign it an analogy with a wave-shock. In any wave shock produce by a blast (big bang for example), there are a zone of high pressure (wave front) and after, other wave or zone of low pressure. This zone, produce (without any drag) one acceleration: next images about wave explosion propagation with simulation CFD techniques (test made with Star CCM+ as a CFD code: simulation in 3D (cut plane view), 10 km diameter sphere, 14.5 million mesh cells, explosion of dynamite into air dry, K-epsilon turbulence model):

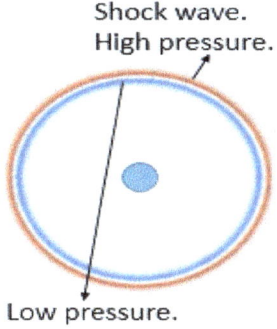

Shock wave.
High pressure.

Low pressure.

The front-shock wave, have only 2 brakes (drag): viscosity and mass (gravity).

It can see the low pressure rear in any car (any wave shock). The pressure profile in any blast wave is (the wave may to be oscillations (positives and negatives) in time). Friedlander waveform sample for any explosion:

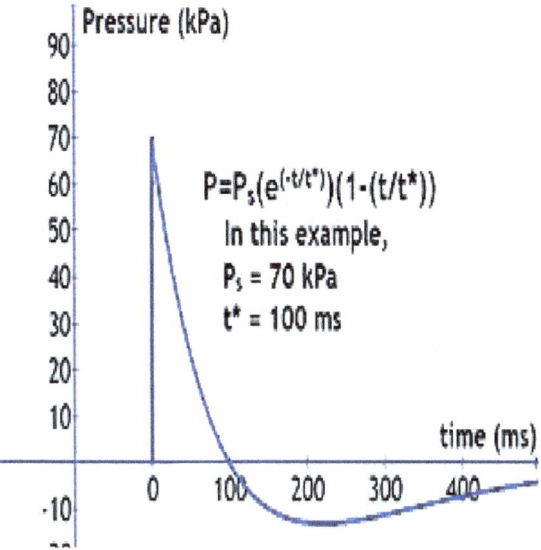

The zone or zones, with negative pressure, the "dark energy", work (the dark energy change in time), producing acceleration-suction (with positive pressure, work but pushing).

Even, combining some hypothesis is possible that in the future, there are variations positives and negatives in dark energy:

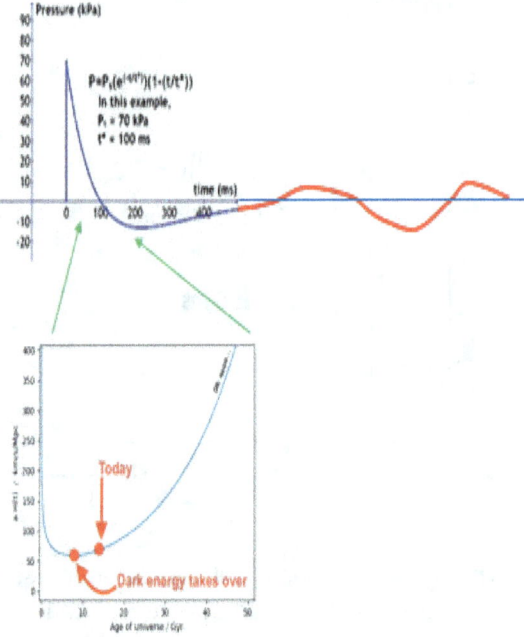

This evolution in waves or not, depend of densities in the universe (baryonic matter, radiation, dark energy, dark matter, electromagnetic, etc).

Applying the Euler Lagrange equations ("E" energy):

$$KineticE = \frac{1}{2}m \overset{*2}{x}$$

$$PotentialE = m\frac{P_x}{\rho}x$$

So: $P_x = \frac{\partial P}{\partial x} = \overset{**}{x}\rho = a\rho$

Navier Stokes expansion universe model

This model of expansion of the Universe has been realized in 1 dimension, assuming that the whole Universe expands equally in any direction. But I don't think this is the case; there have been observations in various directions in the Universe, which show that it is not uniform in all directions (or equal); although there is little data from which to draw final conclusions, I think it is.

For this reason, the expansion of the Universe will depend on pressure and density, so in each direction, the expansion will be different. It works with Navier Stokes equations traditional.

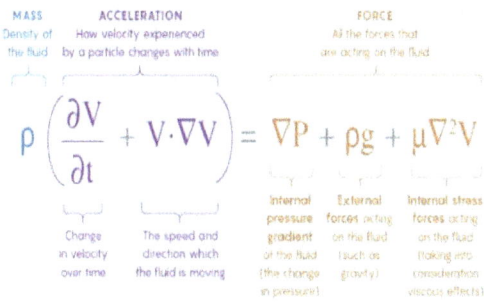

MASS   ACCELERATION
Density of — How velocity experienced
the fluid — by a particle changes with time

FORCE
All the forces that
are acting on the fluid

$$\rho \left(\frac{\partial V}{\partial t} + V \cdot \nabla V \right) = \nabla P + \rho g + \mu \nabla^2 V$$

Change — The speed and
in velocity — direction which
over time — the fluid is moving

Internal — External — Internal stress
pressure — forces acting — forces acting
gradient — on the fluid — on the fluid
of the fluid — (such as — (taking into
(the change — gravity) — consideration
in pressure) — viscous effects)

Navier-Stokes Equations
Describe the flow of incompressible fluids.

Analyzing the expressions for every part in Navier Stokes equations:

Sup: V=Hx

$$\rightarrow \quad \frac{\partial v}{\partial t} = \overset{*}{H} x + H^2 x$$

$$\rightarrow \quad v\frac{\partial v}{\partial x} = H^2 x$$

$$g = \frac{Gm}{x^2} = \frac{4\pi G \rho x}{3}$$

Viscous term:

$$\rightarrow \mu v = \mu H \, x$$

$$\rightarrow \mu \nabla^2 V = H \, x$$

The next term in Navier Stokes, is the most important:

$$\rightarrow \frac{P_x}{\rho} \quad \text{value ????}$$

Method 1:

$$\rho_{vac} = \frac{\Lambda c^2}{8\pi G}$$

$$P_{vac} = -\rho_{vac} c^2$$

$$\Lambda = \frac{3H^2}{c^2} \Omega_\Lambda$$

$$P = -\frac{c^4}{8\pi G} \frac{3H^2}{c^2} \Omega_\Lambda =$$

$$= -\frac{3H^2 c^2}{8\pi G} \Omega_\Lambda$$

Method 2:

It knows also, that:

$$\dot{\rho} + 3\frac{\dot{a}}{a}\left(\rho + \frac{P}{c^2}\right) = 0$$

$$P = \left(\frac{-\overset{*}{\rho}}{3H} - \rho\right)c^2$$

3 results for "P":

→

Derivating the expressions for "P" (in 2 methods).

→   Also: H=H(x), from:

From Navier Stokes, directly:

Sup: V=Hx (particular case):

$$\frac{P_x}{\rho} + g = \frac{P_x}{\rho} + \frac{4\pi G\rho x}{3} = H^2 x$$

$$P_x = \left(H^2 - \frac{4\pi G\rho}{3} \right)\rho x$$

It's easy calculating that, with other 2 models for "H".

It's very interesting, the next expression:

$$P = \frac{1}{2}\rho V^2$$

→ From this special equation, and very common; it has a Navier Stokes equations part:

$$\frac{\partial P}{\partial x} = \rho V \frac{\partial V}{\partial x}$$

If the density is not constant:

$$\frac{\partial P}{\partial x} = \frac{1}{2}\frac{\partial \rho}{\partial x}V^2 + \rho V \frac{\partial V}{\partial x}$$

There is another problem, very important:

If the density or pressure is not the same in any direction or there are different densities in some places, if it calculates (from supernovas, or galaxies cluster or individual galaxies) the "H_0" Hubble constant, the results can will be different.

That is very important, in the famous "Hubble Tension" problem: in the main Friedman equation, the "H" depend of density….

In this case, the density and/or pressure, can be a vector. Even the Viscosity….

- ***Friedman equations expansion universe model***

Method 1:

Kinetic energy (T) + Potential energy "V" = E constant, so:

Sup: V=Hx

$$V=-\left(\frac{4\pi}{3}\rho r^{2}\right)\frac{mG}{r}=-\frac{4}{3}\pi G\rho r^{2} m$$

$$T=\frac{m\overset{*}{r}^{2}}{2} \qquad \text{r=ax}$$

$$E=\frac{m}{2}\overset{*}{r}^{2}\overset{*}{x}^{2}-\frac{4}{3}\pi G m\rho a^{2}x^{2}$$

Dividing by ma^2x^2:

$$\frac{2E}{ma^{2}x^{2}}=\left(\frac{\overset{*}{a}}{a}\right)^{2}-\frac{8\pi}{3}G\rho$$

$$E=\frac{1}{2}mc^{2}x^{2}K$$

$$\left(\frac{\overset{*}{a}}{a}\right)^2 = \frac{8\pi G}{3}\rho - \frac{Kc^2}{a^2} = H^2$$

Method 2.

Applying Navier Stokes equations, with V=H ("g" gravity acceleration):

$$\frac{\partial V}{\partial t} + V\frac{\partial V}{\partial x} = g$$

In order to reach the correct value for critical density of universe, it´s necessary substituting "g" by "4g"; it´s necessary analyze that.... It´s possible may be, that this "4" was necessary in Navier Stoke equations applied to Universe expansion....

Method 3:

From Euler-Lagrange equations: Sup:

$$L = \frac{1}{2}mV^2 + \frac{GmM}{x}$$

$$\frac{\partial L}{\partial x} = \frac{8\pi G\rho}{3}$$

$$\frac{d}{dt}\left(\frac{\partial L}{\partial \overset{*}{x}}\right) = \overset{**}{x}$$

$$H^2 x + \overset{*}{H} x = \frac{8\pi G\rho}{3}$$

"H" constant:

$$\rho = \frac{3 H^2}{8\pi G}$$ Critical Density.

- *Spring-Damper expansion universe model*

Considering that the viscosity of universe works as a damper; also, the mass (gravity) and density so, work as a spring (in this special and particular case, not considering other´s forces).

So ("KS" is a value of spring constant, and "KD" diffusivity of Damper) (is possible that "KS" and "KD" non-constants) (Vacuum(Force)=Fv):

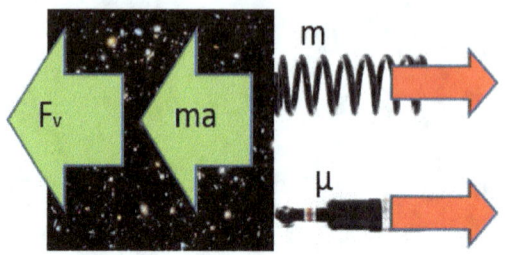

$$F_v - K_S x - K_D \frac{\partial x}{\partial t} = ma$$

About the constants (in general form) ("f" and "g" functions):

$$K_S = f(mass, x) = f(m, x)$$
$$K_D = g(vis\cos ity, x) = g(\mu, x)$$

It´s possible suppose that (reasonable option), is one option:

$$K_S = x$$
$$K_D = Velocity = V$$

Also:

$$K_D \frac{\partial x}{\partial t} = K_D V = F_{viscous}$$
$$K_S x = Force(Gravitational)$$

Also, it´s possible substituting the acceleration "a" in the "Force" (generate by the "full" acceleration) general expression:

$$F = ma = m\frac{\partial V}{\partial x} = m\left(\frac{P_x}{\rho} - g\right) =$$

$$= \overset{*}{H} x + H^2 x$$

$$/ V = H x$$

Very similar to Navier Stokes equations....

Applying the Euler-Lagrange equations:

$$L = \frac{1}{2} m \overset{*}{x}^2 - \frac{1}{2} K_s x^2 + K_D \overset{*}{x}$$

$$\frac{\partial L}{\partial x} = K_s x$$

$$\frac{d}{dt}\left(\frac{\partial L}{\partial \overset{*}{x}}\right) = m \overset{**}{x} + K_s x$$

$$m \overset{**}{x} + K_s x + K_D \overset{*}{x} = 0$$

With Navier Stokes equations and substituting Hubble equation (V=Hx), the importance of "pressure variation" as a dark energy, is more or less the importance of "H²":

$$H^2 \propto \partial P$$

- *Equations and relations important*

The nomenclature and the definitions are the typical in Cosmology.

→ *Friedmann*

$$\left(\frac{\overset{*}{a}}{a}\right)^2 = \frac{8\pi G}{3}\rho - \frac{K c^2}{a^2} = H^2$$

→ *Hubble relation*

$$H = V*x$$

→ *Acceleration (A) expansion*

Non relativistic matter → P=0:

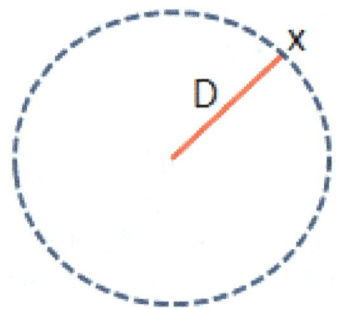

$$V = \overset{*}{a}(t) * R$$

$$A = \overset{**}{a}(t) * R$$

$$D = a(t) * R$$

$$A = -\frac{mG}{D^2} = \overset{**}{a}R$$

$$\overset{**}{a} = \frac{-mG}{a^3 R^3}$$

$$\frac{\overset{**}{a}}{a} = -\frac{4}{3}\pi G\rho$$

Relativistic matter:

$$\left(\frac{\overset{*}{a}}{a}\right)^2 = \frac{8\pi G}{3c^2}\rho - \frac{Kc^2}{a^2 R_0^2}$$

$$\overset{*}{a}{}^2 = \frac{8\pi G}{3c^2}\rho a^2 - \frac{Kc^2}{R_0^2}$$

Deriving:

$$2\,\overset{*}{a}\,\overset{**}{a} = \frac{8\pi G}{3c^2}\left(\rho\,\overset{*}{a}{}^2 + 2\rho a\,\overset{*}{a}\right)$$

With fluid equation:

$$\overset{**}{\frac{a}{a}} = -\frac{4\pi G}{3c^2}\left(\rho + 3P\right)$$

P>0, contributes to negative acceleration; P>0, also contributes to negative acceleration.

→ **Existence (it needs) of Dark Energy Λ**

$$\overset{**}{a} = 0 \leftrightarrow \rho = 0 \rightarrow$$

$$\overset{**}{\frac{a}{a}} = -\frac{4\pi G\rho}{3} + \Lambda$$

→ *Attraction Force between 2 masses*

$$F = \frac{mM}{r^2}G$$

→ *Baryonic matter*

P=0 ; dust: dust gass; good model.

→ *Radiation*

$$P = \frac{\rho c^2}{3}$$

→ *Empty*

Lineal expansion:

$$\frac{\overset{*2}{a}}{a^2} = \frac{-K}{a^2} \to \overset{*}{a} = \sqrt{|K|} \to K < 0$$

➜ **Relation between "H" and "a"**

Matter baryonic:

$$\left(\frac{\overset{*}{a}}{a}\right)^2 = \frac{8}{3}\pi G\rho = \frac{8\pi G}{3}\frac{V}{a^3}$$

$$\left(\frac{\overset{*}{a}}{a}\right)^2 \propto \frac{1}{a^3}$$

Radiation:

$$\left(\frac{\overset{*}{a}}{a}\right)^2 \propto \frac{1}{a^4}$$

➜ **Relation between "a" and "t"**

Only matter:

Method 1:

$$a = c t^p$$

$$\frac{\overset{*}{a}}{a} = \frac{p}{t}$$

$$\left(\frac{\overset{*}{a}}{a}\right)^2 = \frac{p^2}{t^2} = \frac{1}{a^3} = \frac{1}{c^3 t^{3p}} \rightarrow p = 2/3$$

Method 2:

$$\left(\frac{\overset{*}{a}}{a}\right)^2 = \frac{8}{3}\pi G \rho$$

$$\overset{*}{a}{}^2 = \frac{8}{3}\pi G \rho_0 \frac{1}{a}$$

$$\sqrt{a} \, \overset{*}{} \, da = \sqrt{\frac{8}{3}\pi G \rho_0} \, dt \rightarrow t = 0, a = 0 \rightarrow$$

$$\rightarrow \frac{2}{3} a^{3/2} \left(\sqrt{\frac{8}{3}\pi G \rho_0}\right) t$$

$$a(t) = t^{2/3} \sqrt[3]{6\pi G \rho_o}$$

The universe age, is younger so:

$$\frac{\overset{*}{a}}{a} = H = \frac{2}{3}\frac{1}{t} \rightarrow t = \frac{2}{3H}$$

Radiation:

Method 1:

$$\left(\frac{\overset{*}{a}}{a}\right)^2 = \frac{8}{3}\pi G\rho = \frac{8\pi G}{3}\frac{v}{a^4}$$

$$a = ct^p \rightarrow p = 1/2$$

Method 2:

$$E = \sqrt{m^2 + P^2} \approx P, P = \rho/3$$

$$\overset{*}{\rho} + 3\frac{\overset{*}{a}}{a}(\rho + P)$$

$$\overset{*}{\rho} + 4\frac{\overset{*}{a}}{a}\rho = 0$$

$$\overset{*}{\rho} + 3\frac{\overset{*}{a}}{a}(\rho + P)$$

$$\overset{*}{\rho} + 4\frac{\overset{*}{a}}{a}\rho = 0$$

$$\rho = \rho_0\left(\frac{a_0}{a}\right)^4$$

$$\left(\frac{\overset{*}{a}}{a}\right)^2 = \frac{8\pi G}{3}\rho_o \left(\frac{a_0}{a}\right)^4$$

$$\frac{a * da}{a_o^2} = \left(\sqrt{\frac{8\pi G \rho_o}{3}}\right)t$$

$$t = 0, a = 0$$

$$a^2 \propto t; a \propto t^{1/2}$$

$$\frac{a}{a_0} = 2\left(\frac{2\pi G \rho_0}{3}\right)^4 t^{1/2}$$

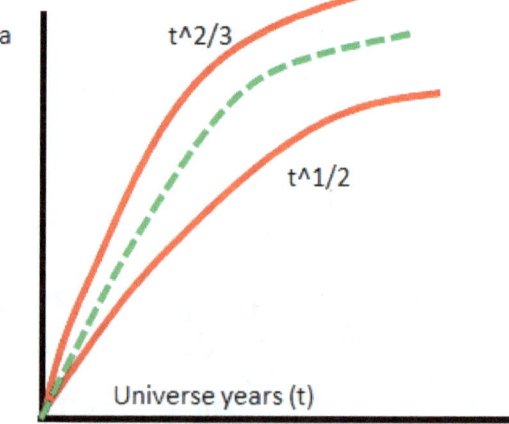

H² total (baryonic + radiation):

$$\left(\frac{\overset{*}{a}}{a}\right)^2 = \frac{C_1}{a^3} + \frac{C_2}{a^4}$$

→ Fluid equation

Variation "Q" =0; internal energy "U(t); density is energy density; "V" volume (sphere), and "P" pressure:

$$(*) \rightarrow dQ = dU + PdV$$
$$U(t) = \rho(t) * V(t)$$

$$U(t) = \frac{4}{3}\pi r^3 a^3 \rho$$

$$dU = 4\pi r^3 a^3 \frac{da}{dt} + \frac{4}{3}\pi r^3 a^3 \frac{d\rho}{dt} =$$

$$= \frac{4}{3}\pi r^3 a^2 \left(3\rho\frac{\overset{*}{a}}{a} + \frac{d\rho}{dt}\right)$$

$$\frac{dV}{dt} = 4\rho r^3 a^2 \frac{da}{dt}$$

Substituting in (*):

$$\frac{4}{3}\pi r^3 a^3 \left(\frac{d\rho}{dt} + 3\frac{\overset{*}{a}}{a}(\rho + P) \right) = 0 \rightarrow$$

$$\rightarrow \frac{d\rho}{dt} + 3\frac{\overset{*}{a}}{a}(\rho + P) = 0$$

$$H = -\frac{\overset{*}{\rho}}{3(\rho + P)}$$

→ Density and Velocity evolution

$$dE = -P * dV$$
$$E = \rho V; dE = \rho dV + V d\rho = -\rho dV$$
$$V d\rho = -(P + \rho)dV; P = w\rho$$
$$V d\rho = -(wP + \rho)dV$$

$$V d\rho = -(w+1)\rho dV$$
$$\frac{d\rho}{\rho} = -(w+1)\frac{dV}{V}$$
$$d\log(\rho) = -(w+1)\log(V)$$
$$\rho = V^{-(w+1)} = \frac{cons\tan t}{V^{w+1}}$$

$$V = a^3$$
$$\rho = \frac{C1}{a^{3(w+1)}}$$

$$w = 0 \rightarrow \rho = \frac{C1}{a^3}; P = 0$$

$$w = 1/3 \rightarrow \rho = \frac{C1}{a^4}; P = \frac{\rho c^2}{3}$$

$$w = -1 \rightarrow \rho = cons \tan t \rightarrow$$

$$\rightarrow H^2 = \frac{8\pi G}{3}\rho_0; P = -\rho c^2$$

$$H = \sqrt{\frac{8\pi G \rho_0}{3}}$$

$$\overset{*}{a} = a\sqrt{\frac{8\pi G \rho_0}{3}}$$

$$a = cons\tan t * e^{\left(\sqrt{\frac{8\pi G \rho_0}{3}}\right)t}$$

$$V = H * D(Hubble) \rightarrow$$

$$\rightarrow V = \left(\sqrt{\frac{8\pi G \rho_0}{3}}\right)D$$

If "V" is "c", then the universe age is between 10-12 billion years.

➜ *Age of Universe*

$$H(a) = H_0 \sqrt{\Omega_m a^{-3} + \Omega_{rad} a^{-3} + \Omega_\Lambda a^{-3}}$$

$$t = \int_0^a \frac{da}{aH(a)}$$

$$V = H * x \rightarrow length = \frac{length}{time} \frac{1}{H}$$

$$so \rightarrow age(today) = \frac{1}{H_0} \approx 13.5By$$

ARTICLE 7

Navier Stokes equations

"P" pressure, "ρ" density and "V" velocity):

$P = \rho V^2$ So, supposing Density constant ($\rightarrow$ now, in 1 Dimension "x"):

$$\frac{\partial P}{\partial x} = \rho 2V \frac{\partial V}{\partial x}$$

The acceleration "acc" is ("t" time):

$$acc = \frac{du}{dt}$$

Sample; if f=f(x,y,z)=f(u,v,w):

$$\frac{df}{dt} = \frac{\partial f}{\partial x}\frac{dx}{dt} + \frac{\partial f}{\partial y}\frac{dy}{dt} + \frac{\partial f}{\partial z}\frac{dz}{dt}$$

$$\frac{dx}{dt} = u; \frac{dy}{dt} = v; \frac{dz}{dt} = w$$

So in 1 dimension "x" (V=V(u)=u):

$$acc = \frac{dV}{dt} = u \frac{\partial u}{\partial x}$$

Also, the units of next expression are acceleration:

$$\frac{\frac{\partial P}{\partial x}}{\rho} = \frac{P_x}{\rho}$$

We can write all the accelerations, as a balance:

$$\frac{\partial u}{\partial t} = -\frac{\frac{\partial P}{\partial x}}{\rho} - u \frac{\partial u}{\partial x} + F(viscous)$$

Obviously, in right terms, is necessary adding the External forces as an electromagnetism, Coriolis effect, gravity, etc....

$$F(viscous) = \frac{\mu}{\rho} \frac{\partial^2 u}{\partial x^2}$$

Mainly nomenclature:

$$\frac{\partial u}{\partial t} + u\frac{\partial u}{\partial x} = -\frac{\dfrac{\partial P}{\partial x}}{\rho} + F(viscous)$$

A   B   C   D

B + C → Euler equation

A +B +D --> Burger equation

A +D → Heat equation

More....

a) ADVECTION LINEAL EQUATION IN 1-D (TRANSPORT WITH VELOCITY "c")

$$\frac{\partial u}{\partial t} + c\frac{\partial u}{\partial x} = 0$$

$$u_i^{n+1} = u_i^n - c\frac{\Delta t}{\Delta x}\left(u_i^n - u_{i-1}^n\right)$$

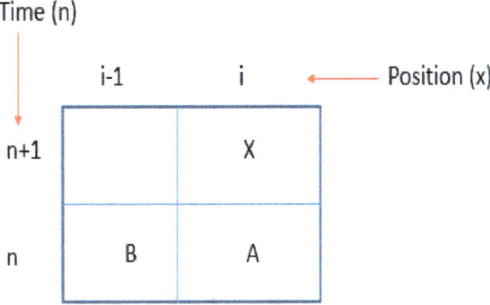

$$X = A - c \frac{\Delta t}{\Delta x}(A - B)$$

- If A>B, then x<A, independently of what scheme work for finites differences (forward, backward, central, etc....).
- X is "A" plus a value, function of a variation (plus or minus).
- If Δt is bigger, the variation is more important (more incorrect) (bigger). That is the basic concept for a inter and extrapolation.
- If Δx is bigger, the variation is smaller.
- "c" is the "risk factor"; if "c" is smaller, the variation is smaller.

Sample (money invests for a "i" and "i-1" people):

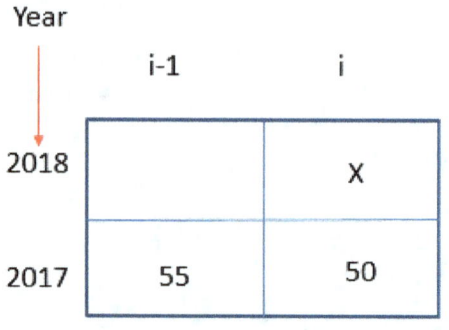

Year

	i-1	i
2018		X
2017	55	50

$$X = 50 - \frac{1}{10}\frac{10}{1}(50-55) = 55$$
$$X = 50 - \frac{1}{10}\frac{10}{5}(50-55) = 50$$

c=0.1

In the first case, "x-1" is a good friend of "x"(Δx=1). In the second case, is a friend not close (Δx=5).

Easy notation (in 1 Dimension; u=u(x,t)):

$$u_t = \frac{\partial u}{\partial t}; u_x = \frac{\partial u}{\partial x}$$

$$Transport - equation:$$

$$u_t + c u_x = 0$$

Generic solution: u(x,t)=f(x-ct)??
Sample to solve:

$$a\frac{\partial u}{\partial x} + b\frac{\partial u}{\partial y} = o$$

$$u(x,y) = v(p,y); p = bx - ay$$

$$\frac{\partial u}{\partial x} = \frac{\partial v}{\partial p}\frac{\partial p}{\partial x} + \frac{\partial p}{\partial y}\frac{\partial y}{\partial x} = b\frac{\partial v}{\partial p}$$

$$\frac{\partial u}{\partial y} = \frac{\partial v}{\partial p}\frac{\partial p}{\partial y} + \frac{\partial v}{\partial y}\frac{\partial y}{\partial y} = -a\frac{\partial v}{\partial p} + \frac{\partial v}{\partial y}$$

Substituting:

$$ab\frac{\partial v}{\partial p} - ba\frac{\partial v}{\partial p} + b\frac{\partial v}{\partial y} = 0 \rightarrow \frac{\partial v}{\partial y} = 0$$

$$\int \frac{\partial v}{\partial y}\,dy = 0$$

$$v(p, y) = f(p)$$

$$u(x, y) = f(bx - ay)$$

Sample solution of ("c=2"), with:

$$u(x,0) = \frac{1}{1 + x^2}$$

So:

$$u(x,t) = \frac{1}{1 + (x - 2t)^2}$$

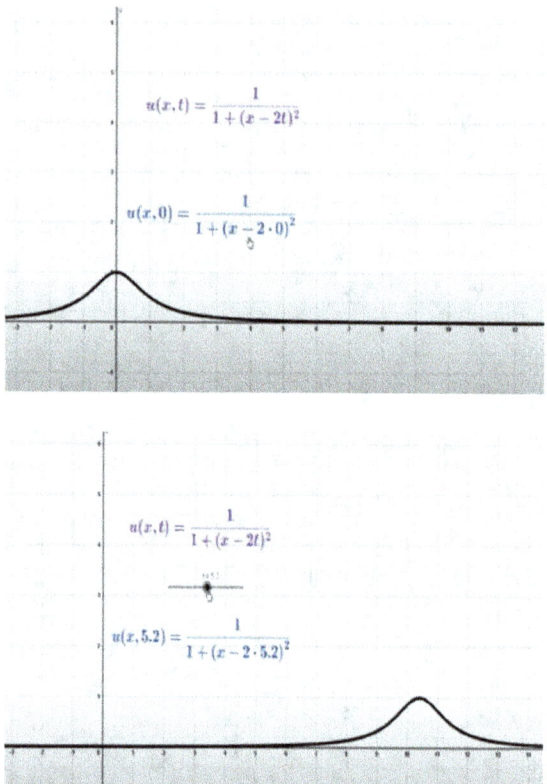

b) ADVECTION LINEAL EQUATION IN 2-D

$$\frac{\partial u}{\partial t} + c\frac{\partial u}{\partial x} + c\frac{\partial u}{\partial y} = 0$$

$$u_{i,j}^{n+1} = u_{i,j}^n - c\frac{\Delta t}{\Delta x}(u_{i,j}^n - u_{i-1,j}^n) - c\frac{\Delta t}{\Delta y}(u_{i,j}^n - u_{i,j-1}^n)$$

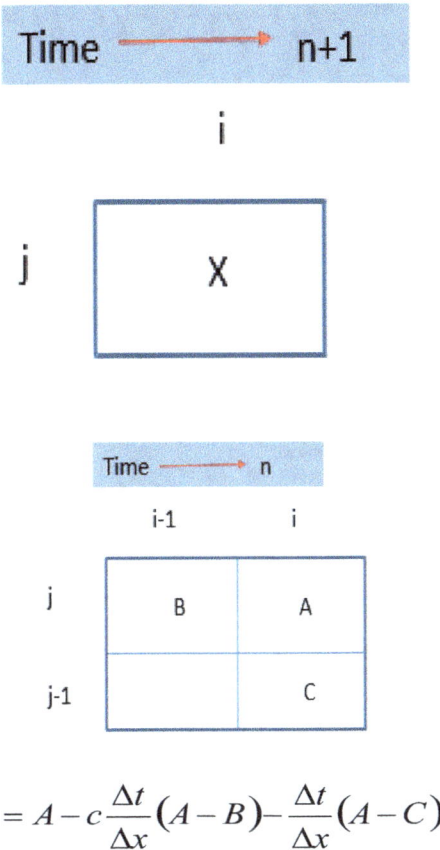

$$X = A - c\frac{\Delta t}{\Delta x}(A - B) - \frac{\Delta t}{\Delta x}(A - C)$$

c) ADVECTION NON LINEAL EQUATION IN 1-D (TRANSPORT WITH VELOCITY "u") → TURBULENCE FORMATION (NONLINEAR EQUATION)

$$\frac{\partial u}{\partial t} + u\frac{\partial u}{\partial x} = 0$$

$$u_x^{n+1} = u_x^n - u_i^n \frac{\Delta t}{\Delta x} \left(u_x^n - u_{x-1}^n \right)$$

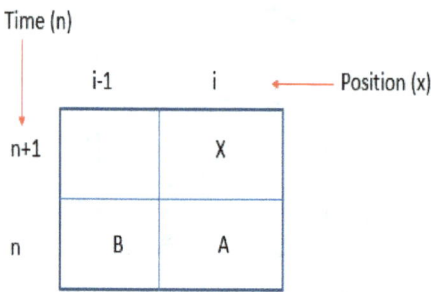

$$X = A - A\frac{\Delta t}{\Delta x}(A - B)$$

This model, allow the turbulence or non-linearity:

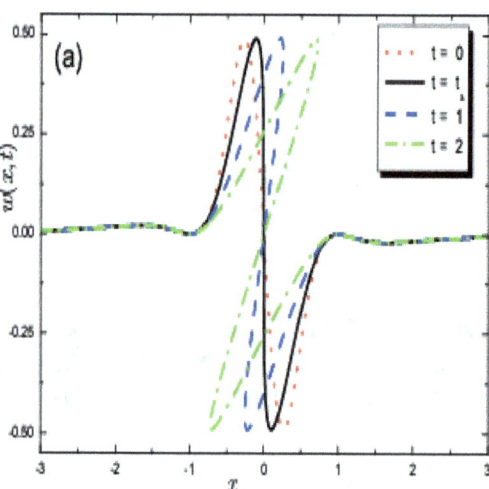

Matlab code and sample, Flow Diffusion using Crank Nicholson:

```
clc
clear
M=100;
N=10;
LX=1;
LY=1;
TIME0=0;
TIME=1;
tt=1000;
Dt=(TIME-TIME0)/tt;
D=12e-4;
DX=LX/M;
DY=LY/N;
mu=D*Dt/(DX)^2;
%Initilization Matrix
for t=1:1;
    for i=2:M-1;
        U(i,t)=10*rand(1,1);
    end
end
%Boundary Conditions
for t=1:1;
U(1,t)=0;
U(M,t)=0;
end
for t=1:1;
    for i=1:1;
        d(i,t)=mu*U(i+1,t)+(1-2*mu)*U(i,t);
    end
end
for t=1:1;
    for i=2:M-1;
        d(i,t)=mu*U(i+1,t)+(1-2*mu)*U(i,t)+mu*U(i-1,t);
    end
end
for t=1:1;
    for i=M:M;
        d(i,t)=(1-2*mu)*U(i,t)+mu*U(i-1,t);
    end
end
%Constructing the Diagonal Matrix
a=ones(M-1,1)
b=ones(M,1)
```

```
g=(1+2*mu)*diag(b)-mu*diag(a,-1)-mu*diag(a,1)
gg=g^-1
for t=1:1;
U(:,t)=gg*d(:,t)
end
for t=1:tt;
   for i=1:1;
        d(i,t)=mu*U(i+1,t)+(1-2*mu)*U(i,t);
   end
   for i=2:M-1;
        d(i,t)=mu*U(i+1,t)+(1-2*mu)*U(i,t)+mu*U(i-1,t);
   end
   for i=M:M;
        d(i,t)=(1-2*mu)*U(i,t)+mu*U(i-1,t);
   end
U(:,t+1)=gg*d(:,t)
end
for t=1:tt;
plot(U(:,t),'-*')
grid on
pause(0.4)
close
end
```

Give a function, to apply the Diffusion equations:

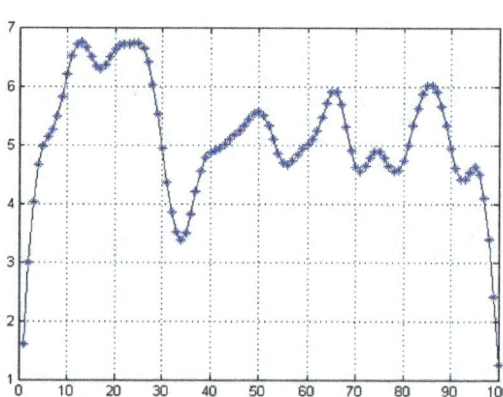

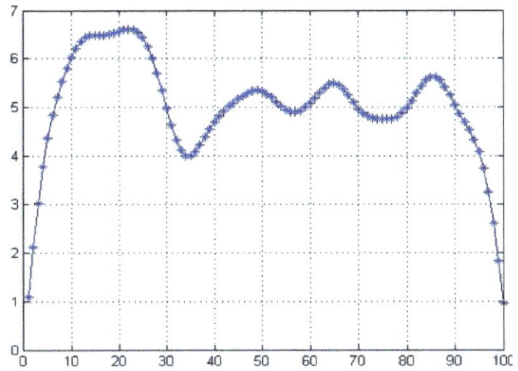

It´s the same case than before, but now, the velocity "c" is a function.

d) ADVECTION NON LINEAL EQUATION IN 2-D

$$\frac{\partial u}{\partial t} + \imath = 0$$

$$\frac{\partial v}{\partial t} + \imath = 0$$

$$u_{i,j}^{n-1} = u_{i,j}^n - \frac{\Delta t}{\Delta x} u_{i,j}^n (u_{i,j}^n - u_{i-1,j}^n) - \frac{\Delta t}{\Delta y} v_{i,j}^n (u_{i,j}^n - u_{i,j-1}^n)$$

$$v_{i,j}^{n-1} = v_{i,j}^n - \frac{\Delta t}{\Delta x} u_{i,j}^n (v_{i,j}^n - v_{i-1,j}^n) - \frac{\Delta t}{\Delta y} v_{i,j}^n (v_{i,j}^n - v_{i,j-1}^n)$$

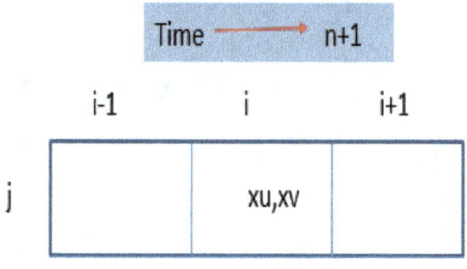

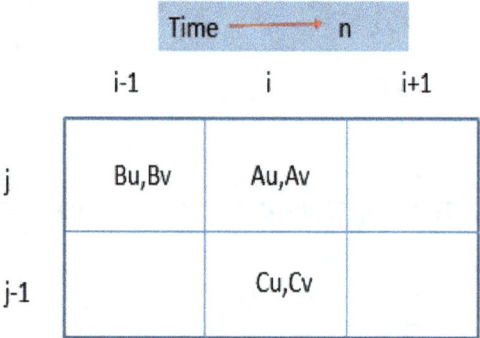

$$xu = Au - \frac{\Delta t}{\Delta x} Au(Au - Bu) - \frac{\Delta t}{\Delta y} Av(Au - Bu)$$

$$xv = Au - \frac{\Delta t}{\Delta x} Au(Av - Bv) - \frac{\Delta t}{\Delta y} Av(Av - Bv)$$

Evolution of a wave, to brake:

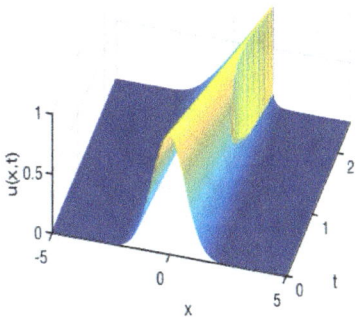

e) LAPLACIAN EXPRESSION IN 1D (heat equation)

"Q" is heat; the variation of "Q" is a heat balance

(entry against exit):

$$\frac{\partial Q}{\partial t} = Q_{int}(x,t)S - Q_{ext}(x,t)S =$$
$$= S\left(Q_i - Q_{ext}\right)$$

m" mass, "u(x,t)" temperature in "x" position and "t" time; "ρ" density, "S" section material, "λ" coefficient transfer heat material:

$$Q(x,t) = \lambda m u(x,t)$$
$$\Delta m = \rho S \Delta x$$

$$Q(x,t) = \lambda m u(x,t) = \lambda \rho S \Delta x \frac{\partial u}{\partial t}$$

$$\lambda \rho S \Delta x \frac{\partial u}{\partial t} = S\left(Q_i - Q_e\right)$$
$$\lambda \rho \frac{\partial u}{\partial t} = \frac{-Q(x+\Delta x,t) - Q(x,t)}{\Delta x}$$

270

$$\lambda \rho \frac{\partial u}{\partial t} = \lim_{\Delta x \to 0} \frac{-Q(x+\Delta x,t)-Q(x,t)}{\Delta x}$$

$$\lambda \rho \frac{\partial u}{\partial t} = -\frac{\partial Q}{\partial x}$$

$$Q(x,t) = -K \frac{\partial Q}{\partial x}$$

The Laplacian (in this case of "u" in 1 Dimension "x"), measures what you could call the « curvature » or *stress* of the field. It tells you how much the value of the field differs from its average value taken over the surrounding points.

The Laplacian "Δf(p)" of a function f at a point "p", up to a constant depending on the dimension, is the rate at which the average value of f over spheres centered at p deviates from "f(p)" as the radius of the sphere grows.

That is, if the Laplacian is positive at a given point, then the average value of the function over a very small sphere centered around that point will be larger than the value of the function at the point. If it's negative, the average will be smaller. If it's zero, the average will be equal. For a harmonic function (everywhere vanishing Laplacian), the function's value always equals the average value over a sphere of any size centered around the point.

If on average over small spheres around is hotter than in then in the next second the temperature in will increases.

→ This tells it that the exchange rate of over time is given by the average rate of change of in space. If it interpret as the temperature (and therefore ∂u/∂t, is the rate of change of temperature), then it can see that there is more heat exchange in regions where the temperature is very variable, and less heat exchange when the temperature varies slightly:

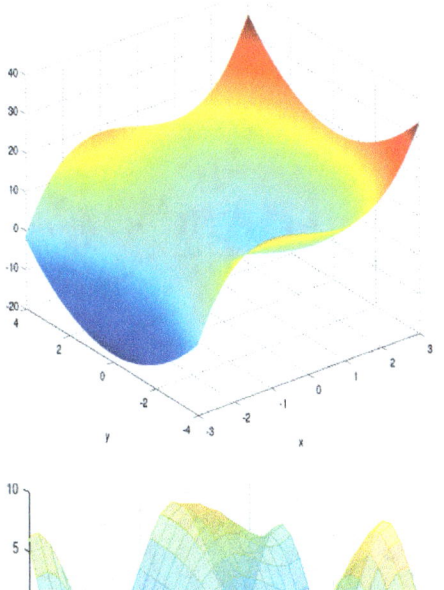

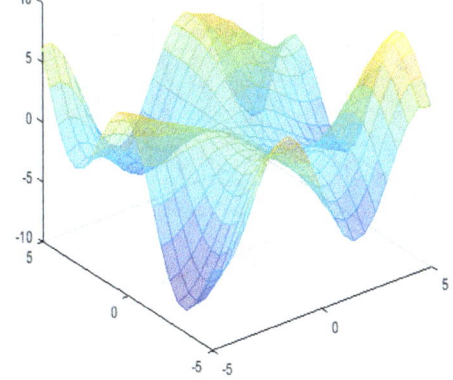

Discretizing the Laplacian expression:

$$u_i^{n+1} = u_i^n + \nu \frac{\Delta t}{\Delta x^2} \left(u_{i+1}^n - 2 u_i^n + u_{i-1}^n \right)$$

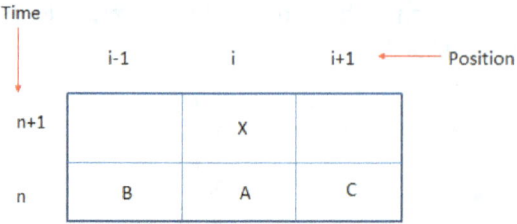

$$x = A + v\frac{\Delta t}{\Delta x^2}(C - 2A + B)$$

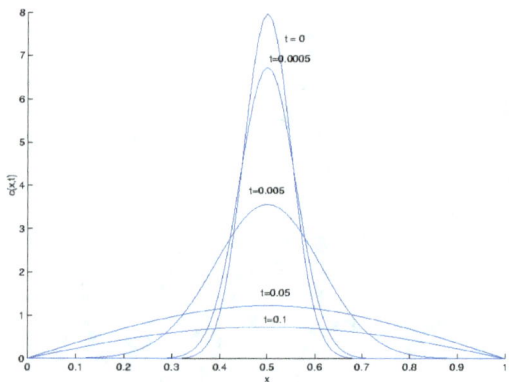

C-2A+B:
- C-2A+B = 0, if the A,B,C in progression lineal (Arithmetic progression).
- >0 if is crescent and <0 in other case.
- Will be a magnitude bigger, when the variation is bigger.
- The lower the "nu", the less heat transfer.
- C-2A+B=(B-A)-(A-C) : that is: variation average between distances in A, B and C.
Sample:
(C,A,B)=(2,8,48)
 2 by 4 = 8
8 by 6 = 48
A-C=6 / B-A=40
(B-A)-(A-C)= 34

This case, it´s a case of heat transfer equation; "K=conductivity/(specific heat*density).

$$u_t = K u_{xx}$$

For example, in one dimension "x", let a bar of "L" as a length; "f(x)" is a heat initial distribution in bar (time=t=0); so, the generic solution is:

$$u(x,t) = \frac{2}{L} \sum_{n=1}^{\infty} \left(\int_0^L f(x) * \sin\left(\frac{n\pi}{L} x\right) dx \right) *$$

$$* e^{-K\left(\frac{n^2 \pi^2}{L^2}\right)t} \sin\left(\frac{n\pi}{L} x\right)$$

f) NAVIER STOKES-BURGER IN 1-D

$$\frac{\partial u}{\partial t} + u \frac{\partial u}{\partial x} = \nu \frac{\partial^2 u}{\partial x^2}$$

$$u_i^{n+1} = u_i^n - u_i^n \frac{\Delta t}{\Delta x}\left(u_i^n - u_{i-1}^n\right) + \nu \frac{\Delta t}{\Delta x^2}\left(u_{i+1}^n - 2u_i^n + u_{i-1}^n\right)$$

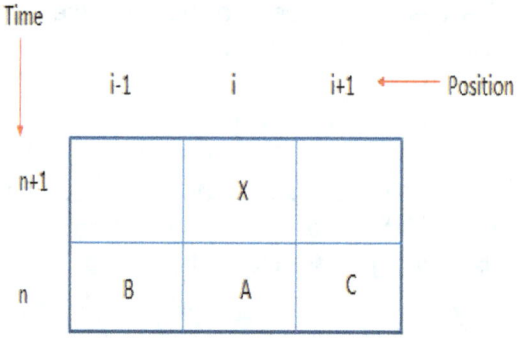

$$X = A - A\frac{\Delta t}{\Delta x}(A - B) + v\frac{\Delta t}{\Delta x^2}(C - 2A + B)$$

g) NAVIER STOKES IN 2D

$$\frac{\partial u}{\partial t} + u\frac{\partial u}{\partial x} + v\frac{\partial u}{\partial y} = -\frac{1}{\rho}\frac{\partial p}{\partial x} + v\left(\frac{\partial^2 u}{\partial x^2} + \frac{\partial^2 u}{\partial y^2}\right)$$

$$\frac{\partial v}{\partial t} + u\frac{\partial v}{\partial x} + v\frac{\partial v}{\partial y} = -\frac{1}{\rho}\frac{\partial p}{\partial y} + v\left(\frac{\partial^2 v}{\partial x^2} + \frac{\partial^2 v}{\partial y^2}\right)$$

The momentum equation in the u direction:

$$u_{i,j}^{n+1} = u_{i,j}^n - u_{i,j}^n\frac{\Delta t}{\Delta x}\left(u_{i,j}^n - u_{i-1,j}^n\right) - v_{i,j}^n\frac{\Delta t}{\Delta y}\left(u_{i,j}^n - u_{i,j-1}^n\right)$$
$$- \frac{\Delta t}{\rho 2\Delta x}\left(p_{i+1,j}^n - p_{i-1,j}^n\right)$$
$$+ v\left(\frac{\Delta t}{\Delta x^2}\left(u_{i+1,j}^n - 2u_{i,j}^n + u_{i-1,j}^n\right) + \frac{\Delta t}{\Delta y^2}\left(u_{i,j+1}^n - 2u_{i,j}^n + u_{i,j-1}^n\right)\right)$$

The momentum equation in the v direction:

$$v_{i,j}^{n+1} = v_{i,j}^n - u_{i,j}^n\frac{\Delta t}{\Delta x}\left(v_{i,j}^n - v_{i-1,j}^n\right) - v_{i,j}^n\frac{\Delta t}{\Delta y}\left(v_{i,j}^n - v_{i,j-1}^n\right))$$
$$- \frac{\Delta t}{\rho 2\Delta y}\left(p_{i,j+1}^n - p_{i,j-1}^n\right)$$
$$+ v\left(\frac{\Delta t}{\Delta x^2}\left(v_{i+1,j}^n - 2v_{i,j}^n + v_{i-1,j}^n\right) + \frac{\Delta t}{\Delta y^2}\left(v_{i,j+1}^n - 2v_{i,j}^n + v_{i,j-1}^n\right)\right)$$

h) WAVE EQUATION IN 1D

From ([45] Jorge Lindley):

Perturbations:

$$u = \epsilon u_1, \quad p = p_0 + \epsilon p_1, \quad \rho = \rho_0 + \epsilon \rho_1$$

Ignoring terms >2 degree:

$$(p_0 + \epsilon p_1)(\rho_0 + \epsilon \rho_1)^{-\gamma} = p_0 \rho_0^{-\gamma},$$

$$\Rightarrow \left(1 + \frac{\epsilon p_1}{p_0}\right)\left(1 + \frac{\epsilon \rho_1}{\rho_0}\right)^{-\gamma} = 1$$

$$\Rightarrow \left(1 + \frac{\epsilon p_1}{p_0}\right)\left(1 - \frac{\gamma \epsilon \rho_1}{\rho_0} + O(\epsilon^2)\right)^{-\gamma} = 1.$$

$$\Rightarrow \frac{p_1}{p_0} = \frac{\gamma \rho_1}{\rho_0}$$

$$\Rightarrow p_1 = c_s^2 \rho_1$$

$$c_s = \sqrt{\frac{\gamma \rho_1}{\rho_0}}.$$

From Navier Stokes linearised:

$$\rho_0 \frac{\partial u_1}{\partial t} = -\nabla p_1$$

$$\frac{\partial \rho_1}{\partial t} + \rho_0 \nabla \cdot u_1 = 0.$$

$$\rho_0 \frac{\partial}{\partial t} \nabla \cdot u_1 = -\Delta p_1$$

Now, it´s very interesting, to know the derivation of wave equation:

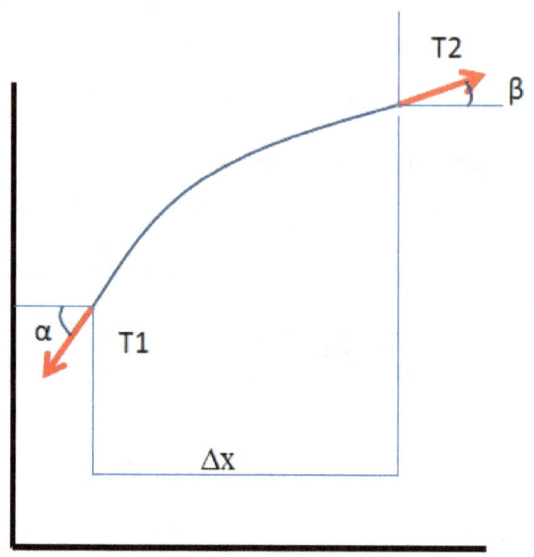

$$T_1 \cos(\alpha) = T_{21} \cos(\beta) = T$$

$$\sum F_{vertical} = m \, a_{vertical}$$

$$\frac{T_2 \sin(\beta)}{T} - \frac{T_1 \sin(\alpha)}{T} = \frac{\rho \Delta x}{T} \frac{\partial^2 u}{\partial t^2}$$

$$\frac{T_2 \sin(\beta)}{T_2 \cos(\beta)} - \frac{T_1 \sin(\alpha)}{T_1 \cos(\alpha)} = ...$$

$$\frac{tg(\beta)}{\Delta x} - \frac{tg(\alpha)}{\Delta x} = ...$$

$$\frac{\left(\dfrac{u}{\partial x}\right)_{x+\Delta x} - \left(\dfrac{u}{\partial x}\right)_x}{\Delta x} = \frac{\rho}{T} \frac{\partial^2 u}{\partial t^2}$$

- **Numerical model with Navier Stokes in more than 3 dimensions**

In the Article ([3] A. Jakimowicza and J. Juzwiszynb), the formation of vortices in the evolution of economic parameters is analyzed; it is appreciated in the image depending on the geometry section (or shadow) that is observed, the helix of the vortex appears; in these type or vortex, normally there is an attractor in this type of problem:

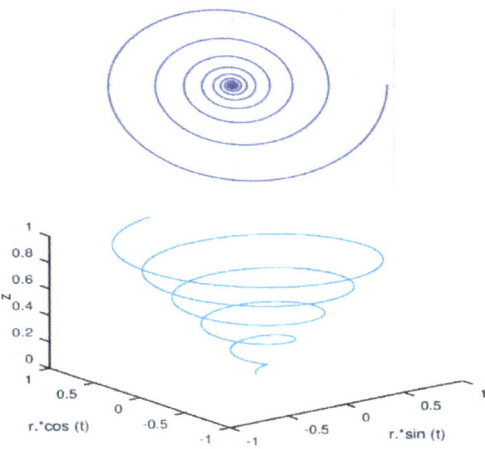

In general, it´s necessary transforming or show the coordinates-geometry in order to analyze attractors or the "hide" geometry:

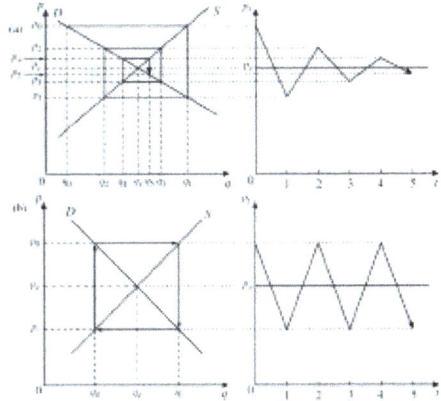

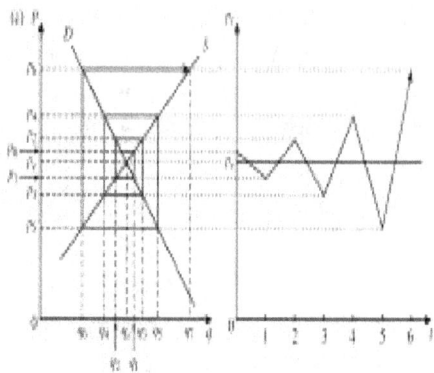

As an example, an event, in 2 coordinates "x" and "y" on which the event depends (the procedure is valid for "n" dimensions):

A Potential is defined which will be incorporated into the term Pressure in the equations of Navier Stokes; this Potential, is the expression (combination of parameters or values) by which the event evolves over time (suction).

The Navier Stokes equations are solved (with specials initials and boundary conditions); you will have a map of pressures (e.g.) in 2 dimensions in which by choosing a point, it will obtain a streamline.

This streamline or path, will be the evolution of the event with respect to time; the potential will be varied to adapt the calculated evolution to the real evolution. If suddenly, there is a factor impossible to determine or know that affects the evolution of the event, the path is recalculated, introducing a new real seed point (and initial-boundary conditions), from which, another path will be obtained. It is also possible to change the potential, to make it more suitable. In both cases, the aim is to improve the model and/or the path. The streamlines, may form spirals or vortices or deviate from a high or low pressure zone/point for example; but the Time, is the third

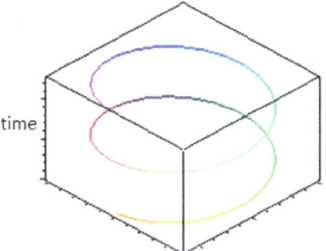

coordinate….:

Depending the seed point, it creates different paths or streamlines (blue: low pressure):

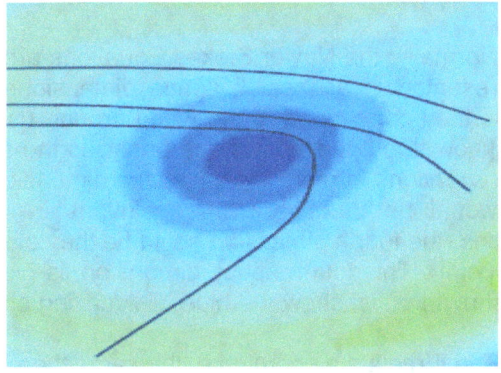

To make the predictions, it is necessary to solve the Navier Stokes Equations in an enclosure of "n" dimensions. It is even possible to place in that space, wall type boundaries; that is to say: objects.

To do this and to correctly define these initial conditions but also boundary conditions, it is necessary to rely on known data and results.

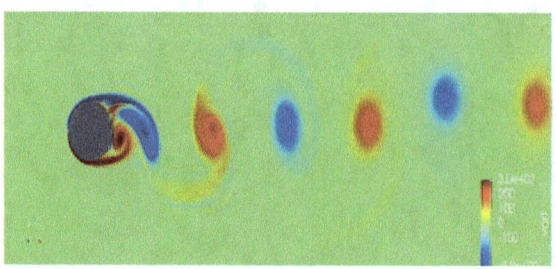

Even, the object in movement….

- *Working with Navier Stokes or any equation in more than 3 space dimensions, even Fractal dimensions*

Normally, the Navier Stokes Equations are studied and represented in at most 3 space dimensions; this is logical as they correspond to the world around us and we need to know it; but it is possible to analyses the solutions of these equations in more than 3 space dimensions; how nice it would be to observe the solutions in 4 space dimensions (the fourth dimension could be the color or size of each point). This fourth space dimension, is hide and it tries to show; that is: show the hide…. Amazing challenge, yes¡¡¡¡

The Time it´s a coordinate, the same than "x"or "y"; not problem about for us, the Mathematicians. So, it´s possible to show, for example in 3 space dimensions: (x,y,z), (x,z,t), (x,y), etc…. If on a 3-coordinate function (x,y,z) we represent e.g. (x,y) or (y,z), it means that the data of a perpendicular or orthogonal section is displayed; that is: planar (flat) sections. But obviously, it´s possible also, work with sections from a curves or even surfaces; that it´s very important.

What mean a section flat with inclination in a 3D space? It´s a representation with scale or deformation in some axis (in this case, between "b" and "a" and "c"); that is:

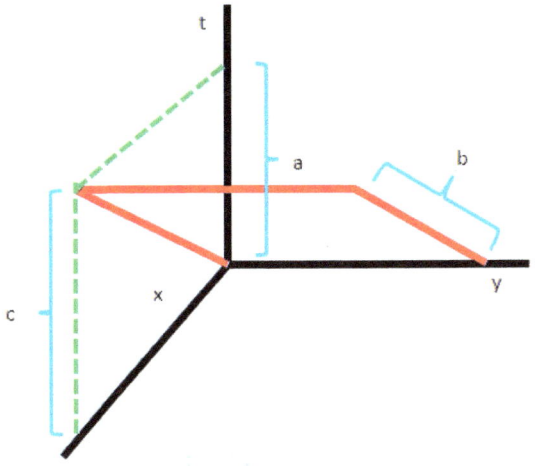

Another important issue is the following: what can these equations look like in a fractal dimension? i.e.; it knows them in 2 and 3 dimensions, but in 2.5 dimensions? It is therefore necessary, first of all, to think how they can be represented or better, what they can mean this fractal dimension. That is: in a 3D geometry, it can do a section planar and will obtain a graphic in 2D:

- How must to do the "section" in order to obtain a graphic in fractal dimension?
- How must to do the "section" in order to obtain a geometry in 3D, from a geometry in 6 dimensions, for example?
- What mean physically the dimensions (more than 3) in Navier Stokes equations?
- What mean physically a fractal dimension?
- There are functions or process for to obtain them, with fractal dimension that it wants?
- There are functions with the same fractal dimension?
- The fourth dimension, touch some edge in 3 dimension?
- How to work with equations in "n" dimensions (also fractal)? That is: with solutions in "n" dimensions. For example: work with wave or heat equation in 5 or 3.5 dimensions....

➔ Many more interesting questions.

- ***Intersections between surfaces; fractals generation***

It´s possible create a fractal geometry, from an intersection between not fractal geometries?

- ***Sections in "other´s" dimensions***

¿¿¿¿

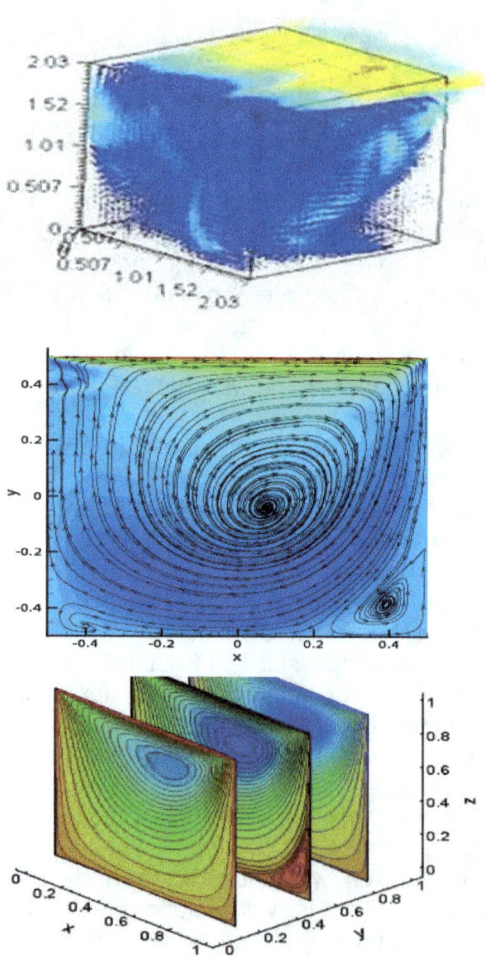

Analyzing Heat Equation in 4 spatials dimensions (+ "t" time): analyzing the "hide" dimensions

From ([44] Ryan C. Daileda) the general solution of 2D Heat Equation is:

$$u_t = c^2 \nabla^2 u = c^2 \left(u_{xx} + u_{yy} \right)$$

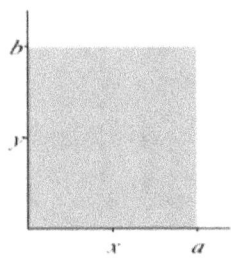

$$u(0, y, t) = u(a, y, t) = 0, \qquad 0 \le y \le b, \ t \ge 0$$
$$u(x, 0, t) = u(x, b, t) = 0, \qquad 0 \le x \le a, \ t \ge 0.$$
$$u(x, y, 0) = f(x, y), \ (x, y) \in R$$

where $R = [0, a] \times [0, b]$.

Assuming that:
$$u(x, y, t) = X(x) Y(y) T(t).$$

Then:

$$X'' - BX = 0, \qquad X(0) = 0, \qquad X(a) = 0.$$
$$Y'' - CY = 0, \qquad Y(0) = 0, \qquad Y(b) = 0.$$
$$T' - c^2 (B + C) T = 0.$$

$$X_m(x) = \sin \mu_m x, \qquad \mu_m = \frac{m\pi}{a}, \qquad B = -\mu_m^2$$

$$Y_n(y) = \sin \nu_n y, \qquad \nu_n = \frac{n\pi}{b}, \qquad C = -\nu_n^2,$$

$$T_{mn}(t) = e^{-\lambda_{mn}^2 t},$$

$$\lambda_{mn} = c\sqrt{\mu_m^2 + \nu_n^2} = c\pi\sqrt{\frac{m^2}{a^2} + \frac{n^2}{b^2}}.$$

$$u_{mn}(x,y,t) = X_m(x)Y_n(y)T_{mn}(t) = \sin \mu_m x \, \sin \nu_n y \, e^{-\lambda_{mn}^2 t}$$

$$u(x,y,t) = \sum_{m=1}^{\infty}\sum_{n=1}^{\infty} A_{mn} \sin \mu_m x \, \sin \nu_n y \, e^{-\lambda_{mn}^2 t}$$

$$f(x,y) = u(x,y,0) = \sum_{n=1}^{\infty}\sum_{m=1}^{\infty} A_{mn} \sin \frac{m\pi}{a} x \, \sin \frac{n\pi}{b} y$$

$$A_{mn} = \frac{4}{ab}\int_0^a\int_0^b f(x,y)\sin\frac{m\pi}{a}x \, \sin\frac{n\pi}{b} \, dy \, dx.$$

Finally, the general solution is:

$$u(x,y,t) = \sum_{m=1}^{\infty}\sum_{n=1}^{\infty} A_{mn} \sin \mu_m x \, \sin \nu_n y \, e^{-\lambda_{mn}^2 t},$$

$$\mu_m = \frac{m\pi}{a}, \; \nu_n = \frac{n\pi}{b}, \; \lambda_{mn} = c\sqrt{\mu_m^2 + \nu_n^2}, \text{ and}$$

$$A_{mn} = \frac{4}{ab}\int_0^a\int_0^b f(x,y)\sin\frac{m\pi}{a}x \, \sin\frac{n\pi}{b} y \, dy \, dx$$

In this point, it´s very interesting analyze the behavior of heat equation in 4 dimensions spatials (x,y,z,w); very interesting so, show the graphic, for example, (x,w), (z,w) or (x,y,w), etc.... and also if exist continuity in heat between dimensions, possible contact or diffusion between edges, etc.... Calculate the solution for a cube of dimensions (a,b,c,d)=(1,1,1,1):

$$u_t = \alpha \left(u_{xx} + u_{yy} + u_{zz} + u_{ww} \right)$$

It´s easy, extend the solution in 2D to 4D (4 summation and 4 integrals). It is assumed that you have the analytical solution of the heat equation in 3 dimensions (x,y,z); it is assumed that from this solution, you solve the heat equation in 4 dimensions (x,y,z,w); the representation of (x,y,z) is the same in both 3D and 4D solutions? In 4D, 3 coordinates are any 3 of the 4 coordinates you have; therefore, if we represent (x,y,w), "w" must coincide with "z" ????....

- **Analyzing Wave Equation in 4 spatials dimensions (+ "t" time): analyzing the "hide" dimensions**

The solution general in 2D, is:

$$\text{Ecuación de onda} \quad \frac{\partial^2 u}{\partial x^2} + \frac{\partial^2 u}{\partial y^2} = \frac{1}{c^2}\frac{\partial^2 u}{\partial t^2} \quad c = \sqrt{\frac{T}{\rho}}$$

$$cf1: u(x,0,t) = u(x,b,t) = 0 \quad ci1: u(x,y,0) = f(x,y)$$

$$cf2: u(0,y,t) = u(a,y,t) = 0 \quad ci2: \frac{\partial u}{\partial t}u(x,y,0) = 0$$

$$\text{solución}: u(x,y,t) = \sum_{m\geq 1}\sum_{n\geq 1} b_{m,n}\phi_{m,n}(x,y)\cos(c\lambda_{m,n}t)$$

$$\phi_{m,n}(x,y) = \sin\left(\frac{m\pi x}{a}\right)\sin\left(\frac{n\pi y}{b}\right) \quad \lambda_{m,n} = \pi\sqrt{\frac{m^2}{a^2}+\frac{n^2}{b^2}}$$

$$b_{m,n} = \frac{4}{ab}\int_0^b\int_0^a \phi_{m,n}(x,y)f(x,y)dxdy$$

As before, is easy applying this expression to 4D.

- **Analyzing Navier Stokes Equation in 4 spatials dimensions (+ "t" time): analyzing the "hide" dimensions**

¿¿¿¿

ARTICLE 8

Schrodinger and Dirac equations

→ *Schrodinger equation*

From ([46] D. Cabrera, P. Fernández de Córdoba, J.M. Isidro, J.M. Valdés Placeres y J. Vázquez Molina):

$$i\hbar\frac{\partial\psi}{\partial t} + \frac{\hbar^2}{2m}\nabla^2\psi - V\psi = 0.$$

It's possible:

$$\psi = \psi_0 \exp\left(S + \frac{i}{\hbar}I\right) = \psi_0 A \exp\left(\frac{i}{\hbar}I\right)$$

$$A=\exp(S)$$

Substituting:

$$\frac{\partial S}{\partial t} + \frac{1}{m}\nabla S \cdot \nabla I + \frac{1}{2m}\nabla^2 I = 0,$$

Real part and imaginary part:

$$\frac{\partial I}{\partial t} + \frac{1}{2m}(\nabla I)^2 + V + U = 0,$$

$$\frac{\partial S}{\partial t} + \frac{1}{m}\nabla S \cdot \nabla I + \frac{1}{2m}\nabla^2 I = 0,$$

$$U := \frac{-\hbar^2}{2m}\frac{\nabla A^2}{A} = \frac{-\hbar^2}{2m}\left[(\nabla S)^2 + \nabla^2 S\right].$$

287

$$\nabla I = m\mathbf{v}_,$$

$$\nabla(v^2) = 2(\mathbf{v} \cdot \nabla)\mathbf{v} + 2\mathbf{v} \times (\nabla \times \mathbf{v})$$

The gradient of real part:

Obtaining the Navier Stokes equations:

$$\frac{\partial \mathbf{v}}{\partial t} + (\mathbf{v} \cdot \nabla)\mathbf{v} + \frac{1}{m}\nabla U + \frac{1}{m}\nabla V = 0,$$

In order to generate the Schrodinger equation: ("KE" kinetic energy, "PE" potential energy, "h" Planck constant, "f" frequency, "m" mass, "v" velocity, "P" movement quantity, "λ" wave length, 1 dimension "x", "t" time:

- **Schrodinger Independent of time**

$$(*) \rightarrow KE + PE = \frac{1}{2}mv^2 + V = E = hf$$

$$P = \frac{h}{\lambda}$$

$$E = \frac{P^2}{2m} + u/P = mv$$

$$\Psi = e^{i(kx-wt)}$$

$$\frac{d\Psi}{dx} = ik\,e^{i(kx-wt)} = ik\Psi$$

$$\frac{d^2\Psi}{dx^2} = (ik)^2 \Psi$$

$$k = \frac{P}{\hbar}; k = \frac{2\pi}{\lambda}; P = \frac{\hbar}{\lambda}; P = \frac{\hbar k}{2\pi}$$

$$\frac{d^2\Psi}{dx^2} = -\left(\frac{P^2}{\hbar^2}\right)\Psi$$

$$-\hbar^2\frac{d^2\Psi}{dx^2} = P\Psi^2$$

Substituting in (*):

$$E\Psi = \frac{P^2}{2m} + V\Psi$$

$$(**) \rightarrow E\Psi = -\hbar^2\frac{d^2\Psi}{dx^2} + V\Psi$$

$$\frac{-\hbar^2}{2m}\nabla^2 + V(r)$$

$$H\Psi(r) = E\Psi(r)$$

- ### Schrodinger Dependent of time

$$E = \hbar w$$

$$\frac{-i}{\hbar}E\Psi = -iw\Psi = \frac{d\Psi}{dt}$$

$$E\Psi = \frac{\hbar}{-i}\frac{d\Psi}{dt} = i\hbar\frac{d\Psi}{dt}$$

Substituting in (**):

$$i\hbar\frac{d\Psi}{dt} = \frac{-\hbar^2}{2m}\frac{d^2\Psi}{dx^2} + V\Psi$$

Solution general Schrodinger equation

$$\Psi(x,t) = \Psi(x)\varphi(t)$$

$$\left(i\hbar\Psi\frac{d\varphi}{dt} = \frac{-\hbar^2}{2m}\frac{d^2\Psi}{dx^2}\varphi + V\Psi\varphi \right)\frac{1}{\Psi\varphi}$$

$$i\hbar\frac{1}{\varphi}\frac{d\varphi}{dt} = \frac{-\hbar^2}{2m}\frac{1}{\Psi}\frac{d^2\Psi}{dx^2}\varphi + V = E$$

$$\frac{d\varphi}{dt} = \frac{-iE}{\hbar}\varphi \rightarrow \int\frac{d\varphi}{\varphi} = \int\frac{-iE}{\hbar}dt$$

$$\ln(\varphi) = \frac{-iE}{\hbar}t + \cos\tan t \rightarrow \varphi = e^{-iKt}$$

Different cases and solutions Schrodinger equation

See now, some different cases, working with operators in order to simplify the notation substituting the derivatives:

Case 1) V(r)=0

$$H = \frac{-\hbar^2}{2m}B^2$$

$$\frac{-\hbar^2}{2m}B^2Y - E\Psi = 0$$

$$\Psi = \Psi_0 e^{iKx}; B = iK; B^2 = -K^2$$

$$\left(\frac{\hbar^2}{2m}K^2 - E\right)\Psi = 0$$

$$\frac{\hbar^2}{2m}K^2 - E = 0$$

$$K = \frac{\pm\sqrt{2mE}}{\hbar}$$

Case 2) Potential step:

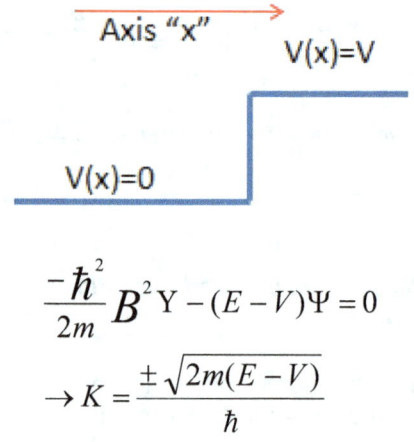

$$\frac{-\hbar^2}{2m}B^2 Y - (E - V)\Psi = 0$$

$$\rightarrow K = \frac{\pm\sqrt{2m(E - V)}}{\hbar}$$

a) E>V:

$$\Psi = \Psi_0 \exp\left(i\frac{\sqrt{2m(E - V)}}{\hbar} x\right)$$

Less frequency and less (E-V):

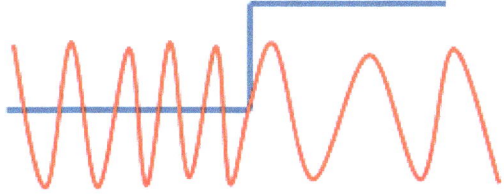

b) E<V:

$$\rightarrow K = \pm i \frac{\sqrt{2m|E-V|}}{\hbar}$$

$$\Psi = \Psi_0 \exp\left(-\frac{\sqrt{2m|E-V|}}{\hbar}x\right) = \Psi_0 \exp(\theta x)$$

The probability "ψ" of find a particle in the material, is less: but not zero ¡¡¡¡:

Material

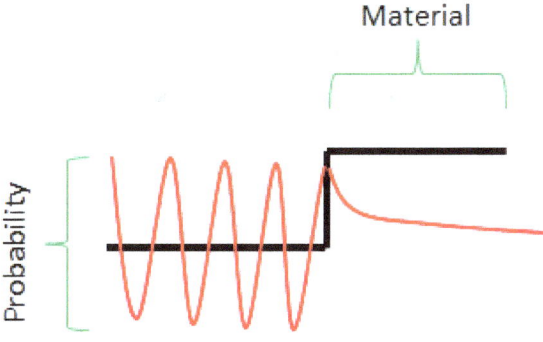

Case 3) Barrier potential: tunnel effect:

The amplitude and the energy "E" are less. Some particle, will go through the material; typical about the radioactivity particles:

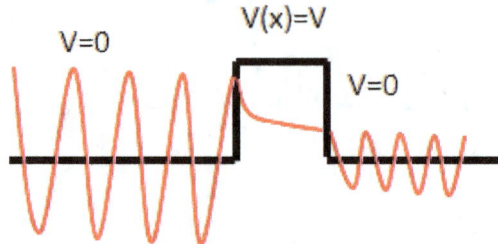

Case 4) Well potential:

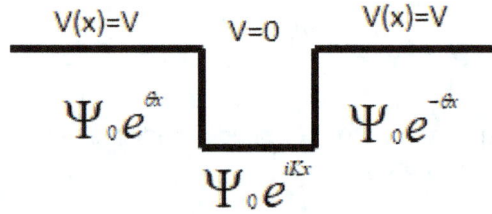

The particles are confined and some, can to exit:

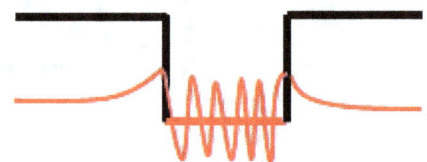

Case 5) Wall potential:
The particle, cannot exit:

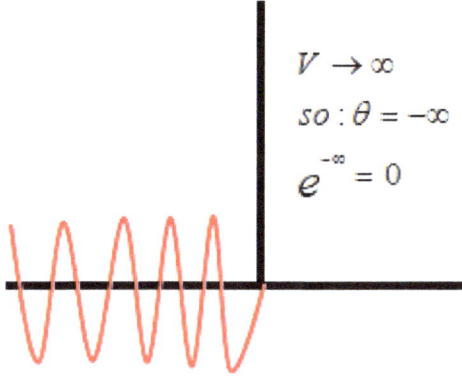

$$V \to \infty$$
$$so : \theta = -\infty$$
$$e^{-\infty} = 0$$

Case 6) Well with walls potential:

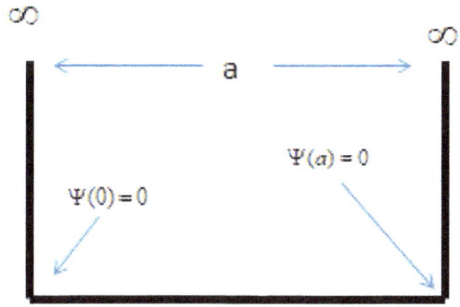

$$\Psi = B_1 e^{iKx} + B_2 e^{-iKx}$$
$$K = \frac{\sqrt{2mE}}{\hbar}$$

The two waves with the same amplitude:

$$\Psi(0) = 0 \to \Psi = B_1 + B_2 = 0$$
$$B_2 = -B_1$$

$$\Psi(x) = B(e^{iKx} - e^{-iKx}) = 2iB\sin(Kx)$$

$$\Psi(a) = 0 \rightarrow \Psi(a) = 2iB\sin(Ka) = 0$$

$$Ka = n\pi / n = 0,1,....$$

$$B = \pm\frac{i}{\sqrt{2a}}$$

$$\Psi(x) = \sqrt{\frac{2}{a}}\sin\left(\frac{n\pi x}{a}\right)$$

It knows that:

$$K = \frac{\sqrt{2mE}}{\hbar} \rightarrow Ka = n\pi$$

$$E = \frac{n^2\pi^2\hbar^2}{2ma^2}$$

- **_Numerical solution Schrodinger equation by Crank Nicholson scheme discretization in 2D_**

$$i\frac{\partial\psi(x,y,t)}{\partial t} = -\nabla^2\psi(x,y,t) + V(x,y,t)\psi(x,y,t)$$

$$i\frac{\partial\psi(x,y,t)}{\partial t} = -\left(\frac{\partial^2\psi(x,y,t)}{\partial x^2} + \frac{\partial^2\psi(x,y,t)}{\partial y^2}\right) + V(x,y,t)\psi(x,y,t)$$

$$\frac{\partial\psi(x,y,t)}{\partial t} = \frac{\psi_{i,j}^{n+1} - \psi_{i,j}^n}{\Delta t}$$

$$\frac{\partial^2\psi(x,y,t)}{\partial x^2} = \frac{1}{2(\Delta x)^2}[(\psi_{i,j+1}^{n+1} - 2\psi_{i,j}^{n+1} + \psi_{i,j-1}^{n+1}) + (\psi_{i,j+1}^n - 2\psi_{i,j}^n + \psi_{i,j-1}^n)]$$

$$\frac{\partial^2\psi(x,y,t)}{\partial y^2} = \frac{1}{2(\Delta y)^2}[(\psi_{i-1,j}^{n+1} - 2\psi_{i,j}^{n+1} + \psi_{i+1,j}^{n+1}) + (\psi_{i-1,j}^n - 2\psi_{i,j}^n + \psi_{i+1,j}^n)]$$

$$V(x,y,z)\psi(x,y,z) = \frac{1}{2}[V_{i,j}^{n+1}\psi_{i,j}^{n+1} + V_{i,j}^n\psi_{i,j}^n]$$

$$r_x = -\frac{\Delta t}{2i(\Delta x)^2}, \quad r_y = -\frac{\Delta t}{2i(\Delta y)^2}$$

$$-r_y\psi_{i+1,j}^{n+1} - r_y\psi_{i-1,j}^{n+1} + (1-2r_x-2r_y+\tfrac{i\Delta t}{2}V_{i,j}^{n+1})\psi_{i,j}^{n+1} - r_x\psi_{i,j+1}^{n+1} - r_x\psi_{i,j-1}^{n+1} =$$

$$= r_y\psi_{i+1,j}^{n} + r_y\psi_{i-1,j}^{n} - (1-2r_x-2r_y-\tfrac{i\Delta t}{2}V_{i,j}^{n})\psi_{i,j}^{n} - r_x\psi_{i,j-1}^{n} + r_x\psi_{i,j-1}^{n}$$

$$a_{ij} = \left(1 + 2r_x + 2r_y - i\frac{\Delta t}{2}V_{i,j}^{n+1}\right)$$

$$b_{ij} = \left(1 - 2r_x - 2r_y - i\frac{\Delta t}{2}V_{i,j}^{n}\right)$$

$$-r_y\psi_{i+1,j}^{n-1} - r_y\psi_{i-1,j}^{n-1} - a_{ij}\psi_{i,j}^{n-1} - r_x\psi_{i,j+1}^{n-1} - r_x\psi_{i,j-1}^{n-1} =$$

$$= r_y\psi_{i+1,j}^{n} + r_y\psi_{i-1,j}^{n} + b_{ij}\psi_{i,j}^{n} - r_x\psi_{i,j-1}^{n} + r_x\psi_{i,j-1}^{n}$$

$$A \cdot x = b$$

$$\psi_{0,j}^{n} = \psi_{i,0}^{n} = \psi_{N-1,j}^{n} = \psi_{i,N-1}^{n} = 0$$

$$-r_y\psi_{i+1,j}^{n-1} - r_y\psi_{i-1,j}^{n-1} - a_{ij}\psi_{i,j}^{n-1} - r_x\psi_{i,j+1}^{n-1} - r_x\psi_{i,j-1}^{n-1} =$$

$$= r_y\psi_{i+1,j}^{n} + r_y\psi_{i-1,j}^{n} + b_{ij}\psi_{i,j}^{n} - r_x\psi_{i,j-1}^{n} + r_x\psi_{i,j-1}^{n}$$

$$A \cdot x =
\begin{pmatrix}
a_{00} & -r_x & 0 & 0 & \cdots & 0 & -r_y & 0 & 0 & \cdots & 0 \\
-r_x & a_{10} & -r_x & 0 & \cdots & & 0 & -r_y & 0 & \cdots & 0 \\
0 & -r_x & a_{20} & -r_x & & & & & -r_y & & \\
0 & 0 & -r_x & \ddots & \ddots & & & & & \ddots & \\
\vdots & \vdots & & \ddots & & & & & & & \\
0 & & & & & & & & & & \\
-r_y & 0 & & & & & & & & & \\
0 & -r_y & & & & & & & & & \\
0 & 0 & -r_y & & & & & & \ddots & & \\
\vdots & \vdots & & \ddots & & & & \ddots & \ddots & -r_x & \\
0 & 0 & & & & & & & -r_x & a_{(N-1),(N-1)} &
\end{pmatrix}
\begin{pmatrix}
\psi_{0,0}^{n-1} \\
\psi_{1,0}^{n-1} \\
\psi_{2,0}^{n-1} \\
\\
\\
\vdots \\
\\
\psi_{i,j}^{n-1} \\
\\
\vdots \\
\psi_{(N-1),(N-1)}^{n-1}
\end{pmatrix}$$

$$k = (j-1)(N-2) - (i-1)$$

$$\psi_{i,j}^{n} = \psi_{k}^{n}$$

$$A \cdot x = \begin{pmatrix} a_0 & -r_x & 0 & 0 & \cdots & 0 & -r_y & 0 & 0 & \cdots & 0 \\ -r_x & a_1 & -r_x & 0 & \cdots & & 0 & -r_y & 0 & \cdots & 0 \\ 0 & -r_x & a_2 & -r_x & & & & & -r_y & & \\ 0 & 0 & -r_x & \ddots & \ddots & & & & & \ddots & \\ \vdots & \vdots & & \ddots & & & & & & & \\ 0 & & & & & & & & & & \\ -r_y & 0 & & & & & & & & & \\ 0 & -r_y & & & & & & & & & \\ 0 & 0 & -r_y & & & & & \ddots & & & \\ \vdots & \vdots & & \ddots & & & & & \ddots & \ddots & -r_x \\ 0 & 0 & & & & & & & & -r_x & a_{(N-3)(N-1)} \end{pmatrix} \begin{pmatrix} \psi_0^{n-1} \\ \psi_1^{n-1} \\ \psi_2^{n-1} \\ \vdots \\ \psi_k^{n-1} \\ \vdots \\ \psi_{(N-3)(N-1)}^{n-1} \end{pmatrix}$$

$$b = M \cdot y = \begin{pmatrix} b_0 & r_x & 0 & 0 & \cdots & 0 & r_y & 0 & 0 & \cdots & 0 \\ r_x & b_1 & r_x & 0 & \cdots & & 0 & r_y & 0 & \cdots & 0 \\ 0 & r_x & b_2 & r_x & & & & & r_y & & \\ 0 & 0 & r_x & \ddots & \ddots & & & & & \ddots & \\ \vdots & \vdots & & \ddots & & & & & & & \\ 0 & & & & & & & & & & \\ r_y & 0 & & & & & & & & & \\ 0 & r_y & & & & & & & & & \\ 0 & 0 & r_y & & & & & \ddots & & & \\ \vdots & \vdots & & \ddots & & & & & \ddots & \ddots & r_z \\ 0 & 0 & & & & & & & & r_x & b_{(N-3)(N-1)} \end{pmatrix} \begin{pmatrix} \psi_0^{n} \\ \psi_1^{n} \\ \psi_2^{n} \\ \vdots \\ \psi_k^{n} \\ \vdots \\ \psi_{(N-3)(N-1)}^{n} \end{pmatrix}$$

- *Numerical solution Schrodinger equation by finite differences scheme discretization in 1D*

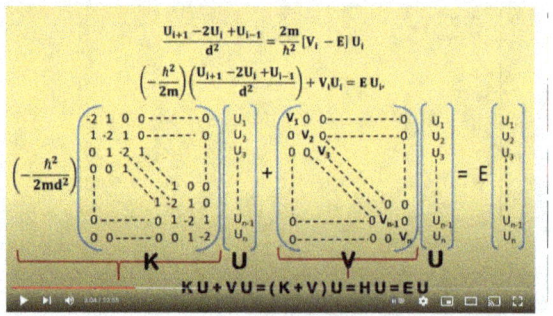

→ *Dirac equation*

For me, one the best and prettiest equation in the world: all, from little idea....

$$\exists a, b / \sqrt{x^2 + b^2} = ax + by \rightarrow$$
$$\rightarrow a^2 = b^2 = 0; ab + ba = 0$$

That it's impossible with Number Real; but not with Matrices (Dirac idea ¡¡¡¡); with 1, 2 and 3 dimensions, not exist solution for this problem, but yes in 4 dimensions (there are others solutions, but this is typical):

$$a = \begin{pmatrix} 0 & 0 & 0 & 1 \\ 0 & 0 & 1 & 0 \\ 0 & 1 & 0 & 0 \\ 1 & 0 & 0 & 0 \end{pmatrix} b = \begin{pmatrix} 1 & 0 & 0 & 0 \\ 0 & 1 & 0 & 0 \\ 0 & 0 & -1 & 0 \\ 0 & 0 & 0 & -1 \end{pmatrix}$$

$$Hamilton \rightarrow E = \frac{P^2}{2m} \rightarrow$$

$$Schrodinger \rightarrow \left(i\hbar \frac{\partial}{\partial t} \right) \Psi = \frac{1}{2m} \left(-i\hbar \frac{\partial}{\partial x} \right) \Psi$$

$$Einstein \rightarrow E = \sqrt{(pc)^2 + (mc^2)^2}$$

If:

$$P = i\hbar \frac{\partial}{\partial x}; E = i\hbar \frac{\partial}{\partial t}$$

Can originate problems with a double derivative (P^2) inside a square root: that, not a meaning physical; so, the Dirac idea in this case is; exist "a" and "b"? :

$$E = aPc + bm c^2$$

Yes ¡¡¡¡:

$$\left(i\hbar \frac{\partial}{\partial t} \right) \Psi = a \left(i\hbar c \frac{\partial}{\partial x} \right) \Psi + bm c^2 \Psi$$

Substituting:

$$E \begin{pmatrix} n1 \\ n2 \\ n3 \\ n4 \end{pmatrix} = \begin{pmatrix} mc^2 & 0 & 0 & Pc \\ 0 & mc^2 & Pc & 0 \\ 0 & Pc & -mc^2 & 0 \\ Pc & 0 & 0 & -mc^2 \end{pmatrix} \begin{pmatrix} n1 \\ n2 \\ n3 \\ n4 \end{pmatrix}$$

The solutions for this system are the values of Energy; for that, it´s necessary to calculate the eigen values:

$$E = \pm\sqrt{(Pc)^2 + m^2 c^4}$$

Are 2 values for the Energy; that is: 2 values for a particle: positive and negative; Energy negative ???? → 2 components (spin).

This expression, it´s the origin for the antimatter....; the Empty, for Dirac, it´s not empty: there are a lot particles with Energy negative (Dirac sea):

- The not existence of E negative, produce E positive.
- The not existence of charge negative, produce charge positive.